HOC SOLUM SCIO QUOD NIHIL SCIO

SCIENCE

INQUIRIES INTO HUMAN
FACULTY AND ITS DEVELOPMENT

AMS PRESS

NEW YORK

INQUIRIES *into*
HUMAN FACULTY
AND ITS
DEVELOPMENT
by Francis GALTON
F·R·S

LONDON: PUBLISHED
by J·M·DENT·&·SONS·L^{TD}
AND IN NEW YORK
BY E·P·DUTTON&CO

Library of Congress Cataloging in Publication Data

Galton, Sir Francis, 1822-1911.
  Inquiries into human faculty and its development.

  Reprint of the 1908 ed., which was issued in series:
Everyman's library.
  Bibliography: p.
  1. Ability. 2. Eugenics. I. Title.
BF701.G23 1973   150   72-1638
ISBN 0-404-08127-4

Reprinted from the edition of 1907, London and New York
First AMS edition published in 1973
Manufactured in the United States of America

AMS PRESS INC.
NEW YORK, N. Y.   10003

PREFACE TO THE SECOND EDITION

AFTER some years had passed subsequent to the publication of this book in 1883, its publishers, Messrs. Macmillan, informed me that the demand for it just, but only just warranted a revised issue. I shrank from the great trouble of bringing it up to date because it, or rather many of my memoirs out of which it was built up, had become starting-points for elaborate investigations both in England and in America, to which it would be difficult and very laborious to do justice in a brief compass. So the question of a Second Edition was then entirely dropped. Since that time the book has by no means ceased to live, for it continues to be quoted from and sought for, but is obtainable only with difficulty, and at much more than its original cost, at sales of second-hand books. Moreover, it became the starting-point of that recent movement in favour of National Eugenics (see note p. 24 in first edition) which is recognised by the University of London, and has its home in University College.

Having received a proposal to republish the book in its present convenient and inexpensive form, I gladly accepted it, having first sought and received an obliging assurance from Messrs. Macmillan that they would waive all their claims to the contrary in my favour.

The following small changes are made in this edition. The illustrations are for the most part reduced in size to suit the smaller form of the volume, the lettering of the composites is rearranged, and the coloured illustration is reproduced as closely as circumstances permit. Two chapters are omitted, on "Theocratic Intervention" and on the "Objective Efficacy of Prayer." The earlier part of the latter was too much abbreviated from the original memoir in the *Fortnightly Review*, 1872, and gives, as I now perceive, a somewhat inexact impression of its object, which was to investigate certain views then thought orthodox, but which are growing obsolete. I could not reinsert these omissions now with advantage, unless considerable additions were made to the references, thus giving more appearance of personal controversy to the memoirs than is desirable. After all, the omission of these two chapters, in which I find nothing to recant, improves, as I am told, the general balance of the book.   FRANCIS GALTON.

LIST OF WORKS.

The Telotype : a printing Electric Telegraph, 1850 ; The Narrative of an Explorer in Tropical South Africa, 1853, in " Minerva Library of Famous Books," 1889 ; Notes on Modern Geography (Cambridge Essays, 1855, etc.); Arts of Campaigning : an Inaugural Lecture delivered at Aldershot, 1855 ; The Art of Travel, or Shifts and Contrivances available in Wild Countries, 1855, 1856, 1860 (1859) ; fourth edition, recast and enlarged, 1867, 1872 ; Vacation Tourists and Notes on Travel, 1861, 1862, 1864 ; Meteorographica, or Methods of Mapping the Weather, 1863 ; Hereditary Genius : an Enquiry into its Laws and Consequences, 1869 ; English Men of Science : their Nature and Nurture, 1874 : Address to the Anthropological Departments of the British Association (Plymouth, 1877) ; Generic Images : with Autotype Illustrations (from the Proceedings of the Royal Institution), 1879 ; Inquiries into Human Faculty and its Development, 1883 ; Record of Family Faculties, 1884 ; Natural Inheritance, 1889 ; Finger-Prints, 1892 ; Decipherments of Blurred Finger-Prints (supplementary chapters to former work), 1893 ; Finger-Print Directories, 1895 ; Introduction to Life of W. Cotton Oswell, 1900 ; Index to Achievements of Near Kinsfolk of some of the Fellows of the Royal Society, 1904 ; Eugenics : its Definition, Scope, and Aims (Sociological Society Papers, vols. I. and II.), 1905 ; Noteworthy Families (Modern Science) ; And many papers in the Proceedings of the Royal Society, Journals of the Geographical Society and the Anthropological Institute, the Reports of the British Association, the Philosophical Magazine, and Nature.

Galton also edited Hints to Travellers, 1878 ; Life-History Album (British Medical Association), 1884, second edition, 1902 ; Biometrika (edited in consultation with F. G. and W. F. R. Weldon), 1901, etc. ; and under his direction was designed a Descriptive List of Anthropometric Apparatus, etc., 1887.

LIST OF MEMOIRS.

The following Memoirs by the author have been freely made use of in the following pages :—

1863 : The First Steps towards the Domestication of Animals (*Journal of Ethnological Society*); 1871 : Gregariousness in Cattle and in Men (*Macmillan's Magazine*); 1872 : Statistical Inquiries into the Efficacy of Prayer (*Fortnightly Review*); 1873 : Relative Supplies from Town and Country Families to the Population of

Future Generations (*Journal of Statistical Society*); Hereditary Improvement (*Fraser's Magazine*); Africa for the Chinese (*Times*, June 6); 1875: Statistics by Intercomparison (*Philosophical Magazine*); Twins, as a Criterion of the Relative Power of Nature and Nurture (*Fraser's Magazine*, and *Journal of Anthropological Institute*); 1876: Whistles for Determining the Upper Limits of Audible Sound (*S. Kensington Conferences*, in connection with the Loan Exhibition of Scientific Instruments, p. 61); 1877: Presidential Address to the Anthropological Department of the British Association at Plymouth (*Report of British Association*); 1878: Composite Portraits (*Nature*, May 23, and *Journal of Anthropological Institute*); 1879: Psychometric Experiments (*Nineteenth Century*, and *Brain*, part vi.); Generic Images (*Nineteenth Century*; *Proceedings of Royal Institution*, with plates); Geometric Mean in Vital and Social Statistics (*Proceedings of Royal Society*); 1880: Visualised Numerals (*Nature*, Jan. 15 and March 25, and *Journal of Anthropological Institute*); Mental Imagery (*Fortnightly Review*; *Mind*); 1881: Visions of Sane Persons (*Fortnightly Review*; and *Proceedings of Royal Institution*); Composite Portraiture (*Journal of Photographical Society of Great Britain*, June 24); 1882: Physiognomy of Phthisis (*Guy's Hospital Reports*, vol. xxv.); Photographic Chronicles from Childhood to Age (*Fortnightly Review*); The Anthropometric Laboratory (*Fortnightly Review*); 1883: Some Apparatus for Testing the Delicacy of the Muscular and other Senses (*Journal of Anthropological Institute*, 1883, etc.).

Memoirs in Eugenics.

1901: Huxley Lecture, Anthropological Institute (*Nature*, Nov. 1901); Smithsonian Report for 1901 (*Washington*, p. 523); 1904: Eugenics, its Definition, Scope and Aims (Sociological Paper, vol. i., *Sociological Institute*); 1905: Restrictions in Marriage, Studies in National Eugenics, Eugenics as a Factor in Religion (Sociological Papers, vol. ii.); 1907: Herbert Spencer Lecture, University of Oxford, on Probability the Foundation of Eugenics.

The following books by the author have been referred or alluded to in the following pages :—

1853: Narrative of an Explorer in Tropical South-Western Africa (*Murray*); 1854: Art of Travel (several subsequent editions, the last in 1872, *Murray*); 1869: Hereditary Genius, its Laws and Consequences (*Macmillan*); 1874: English Men of Science, their Nature and their Nurture (*Macmillan*).

CONTENTS

Contents

Contents

Contents

xviii Contents

PLATES

INQUIRIES INTO
HUMAN FACULTY

INTRODUCTION.

SINCE the publication of my work on *Hereditary Genius* in 1869, I have written numerous memoirs, of which a list is given in an earlier page, and which are scattered in various publications. They may have appeared desultory when read in the order in which they appeared, but as they had an underlying connection it seems worth while to bring their substance together in logical sequence into a single volume. I have revised, condensed, largely re-written, transposed old matter, and interpolated much that is new; but traces of the fragmentary origin of the work still remain, and I do not regret them. They serve to show that the book is intended to be suggestive, and renounces all claim to be encyclopedic. I have indeed, with that object, avoided going into details in not a few cases where I should otherwise have written with fulness, especially in the Anthropometric part. My general object has been to take note of the varied hereditary faculties of different men, and of the great differences in different families and races, to learn how far history may have shown the practicability of supplanting inefficient human stock by better strains, and to consider whether it might not be our duty to do so by such efforts as may be reasonable, thus exerting ourselves to further the ends of evolution more rapidly and with less distress than if events were left to their own course. The subject is, however, so entangled with collateral considerations that a straightforward step-by-step inquiry did not seem to be the most suitable course. I thought it safer to proceed like the surveyor of a new country, and endeavour to fix in the first instance as truly as I could the position of several cardinal points. The general outline of the results to which I finally arrived became more coherent and clear as this process went on; they are briefly summarised in the concluding chapter.

Variety of Human Nature.

We must free our minds of a great deal of prejudice before we can rightly judge of the direction in which different races need to be improved. We must be on our guard against taking our own instincts of what is best and most seemly, as a criterion for the rest of mankind. The instincts and faculties of different men and races differ in a variety of ways almost as profoundly as those of animals in different cages of the Zoological Gardens; and however diverse and antagonistic they are, each may be good of its kind. It is obviously so in brutes; the monkey may have a horror at the sight of a snake, and a repugnance to its ways, but a snake is just as perfect an animal as a monkey. The living world does not consist of a repetition of similar elements, but of an endless variety of them, that have grown, body and soul, through selective influences into close adaptation to their contemporaries, and to the physical circumstances of the localities they inhabit. The moral and intellectual wealth of a nation largely consists in the multifarious variety of the gifts of the men who compose it, and it would be the very reverse of improvement to make all its members assimilate to a common type. However, in every race of domesticated animals, and especially in the rapidly-changing race of man, there are elements, some ancestral and others the result of degeneration, that are of little or no value, or are positively harmful. We may, of course, be mistaken about some few of these, and shall find in our fuller knowledge that they subserve the public good in some indirect manner; but, notwithstanding this possibility, we are justified in roundly asserting that the natural characteristics of every human race admit of large improvement in many directions easy to specify.

I do not, however, offer a list of these, but shall confine myself to directing attention to a very few hereditary characteristics of a marked kind, some of which are most desirable and others greatly the reverse; I shall also describe new methods of appraising and defining them. Later on in the book I shall endeavour to define the place and duty of man in the furtherance of the great scheme of evolution, and

I shall show that he has already not only adapted circumstance to race, but also, in some degree and often unconsciously, race to circumstance; and that his unused powers in the latter direction are more considerable than might have been thought.

It is with the innate moral and intellectual faculties that the book is chiefly concerned, but they are so closely bound up with the physical ones that these must be considered as well. It is, moreover, convenient to take them the first, so I will begin with the features.

FEATURES.

The differences in human features must be reckoned great, inasmuch as they enable us to distinguish a single known face among those of thousands of strangers, though they are mostly too minute for measurement. At the same time, they are exceedingly numerous. The general expression of a face is the sum of a multitude of small details, which are viewed in such rapid succession that we seem to perceive them all at a single glance. If any one of them disagrees with the recollected traits of a known face, the eye is quick at observing it, and it dwells upon the difference. One small discordance overweighs a multitude of similarities and suggests a general unlikeness; just as a single syllable in a sentence pronounced with a foreign accent makes one cease to look upon the speaker as a countryman. If the first rough sketch of a portrait be correct so far as it goes, it may be pronounced an excellent likeness; but a rough sketch does not go far; it contains but few traits for comparison with the original. It is a suggestion, not a likeness; it must be coloured and shaded with many touches before it can really resemble the face, and whilst this is being done the maintenance of the likeness is imperilled at every step. I lately watched an able artist painting a portrait, and endeavoured to estimate the number of strokes with his brush, every one of which was thoughtfully and firmly given. During fifteen sittings of three working hours each—that is to say, during forty-five hours, or two thousand four hundred minutes—he worked at the average rate of ten strokes of the brush per minute. There were, therefore, twenty-four thousand separate traits in the completed portrait, and in his opinion some, I do not say equal, but comparably large number of units of resemblance with the original.

The physiognomical difference between different men being so numerous and small, it is impossible to measure and compare them each to each, and to discover by ordinary statistical methods the true physiognomy of a race. The usual way is to select individuals who are judged to be representatives of the prevalent type, and to photograph them ; but this method is not trustworthy, because the judgment itself is fallacious. It is swayed by exceptional and grotesque features more than by ordinary ones, and the portraits supposed to be typical are likely to be caricatures. One fine Sunday afternoon I sat with a friend by the walk in Kensington Gardens that leads to the bridge, and which on such occasions is thronged by promenaders. It was agreed between us that whichever first caught sight of a typical John Bull should call the attention of the other. We sat and watched keenly for many minutes, but neither of us found occasion to utter a word.

The prevalent type of English face has greatly changed at different periods, for after making large allowance for the fashion in portrait painting of the day, there remains a great difference between the proportion in which certain casts of features are to be met with at different dates. I have spent some time in studying the photographs of the various portraits of English worthies that have been exhibited at successive loan collections, or which are now in the National Portrait Gallery, and have traced what appear to be indisputable signs of one predominant type of face supplanting another. For instance, the features of the men painted by and about the time of Holbein have usually high cheek-bones, long upper lips, thin eyebrows, and lank dark hair. It would be impossible, I think, for the majority of modern Englishmen so to dress themselves and clip and arrange their hair, as to look like the majority of these portraits.

Englishmen are now a fair and reddish race, as may be seen from the Diagram, taken from the Report of the Anthropometric Committee to the British Association in 1880 and which gives the proportion in which the various colours of hair are found among our professional classes.

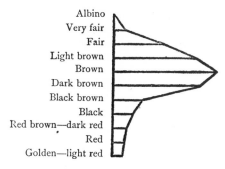

Albino
Very fair
Fair
Light brown
Brown
Dark brown
Black brown
Black
Red brown—dark red
Red
Golden—light red

I take the professional classes because they correspond with the class of English worthies better than any of the others from which returns have been collected. The Diagram, however, gives a fair representation of other classes of the community. For instance, I have analysed the official records of the very carefully-selected crews of H.M.S. *Alert* and *Discovery* in the Arctic Expedition of 1875–6, and find the proportion of various shades of hair to be the same among them as is shown in the Diagram. Seven-tenths of the crews had complexions described as light, fair, fresh, ruddy or freckled, and the same proportion had blue or gray eyes. They would have contrasted strongly with Cromwell's regiment of Ironsides, who were recruited from the dark-haired men of the fen districts, and who are said to have left the impression on contemporary observers as being men of a peculiar breed. They would also probably have contrasted with any body of thoroughgoing Puritan soldiers taken at haphazard; for there is a prevalence of dark hair among men of atrabilious and sour temperament.

If we may believe caricaturists, the fleshiness and obesity of many English men and women in the earlier years of this century must have been prodigious. It testifies to the grosser conditions of life in those days, and makes it improbable that the types best adapted to prevail then would be the best adapted to prevail now.

COMPOSITE PORTRAITURE.

As a means of getting over the difficulty of procuring really representative faces, I contrived the method of composite portraiture, which has been explained of late on many occasions, and of which a full account will be found in Appendix A. The principle on which the composites are made will best be understood by a description of my earlier and now discarded method; it was this—(1) I collected photographic portraits of different persons, all of whom had been photographed in the same aspect (say full face), and under the same conditions of light and shade (say with the light coming from the right side). (2) I reduced their portraits photographically to the same size, being guided as to scale by the distance between any two convenient points of reference in the features; for example, by the vertical distance between two parallel lines, one of which passed through the middle of the pupils of the eyes and the other between the lips. (3) I superimposed the portraits like the successive leaves of a book, so that the features of each portrait lay as exactly as the case admitted, in front of those of the one behind it, eye in front of eye and mouth in front of mouth. This I did by holding them successively to the light and adjusting them, then by fastening each to the preceding one with a strip of gummed paper along one of the edges. Thus I obtained a book, each page of which contained a separate portrait, and all the portraits lay exactly in front of one another. (4) I fastened the book against the wall in such a way that I could turn over the pages in succession, leaving in turn each portrait flat and fully exposed. (5) I focused my camera on the book fixed it firmly, and put a sensitive plate inside it. (6) I began photographing, taking one page after the other in succession without moving the camera, but putting on the cap whilst I was turning over the pages, so that an image of each of the portraits in succession was thrown on the same part of the sensitised plate.

Only a fraction of the exposure required to make a good picture was allowed to each portrait. Suppose that period was twenty seconds, and that there were ten portraits, then an exposure of two seconds would be allowed for each portrait, making twenty seconds in all. This is the principle of the process, the details of that which I now use are different and complex. They are fully explained in the Appendix for the use of those who may care to know about them.

The effect of composite portraiture is to bring into evidence all the traits in which there is agreement, and to leave but a ghost of a trace of individual peculiarities. There are so many traits in common in all faces that the composite picture when made from many components is far from being a blur; it has altogether the look of an ideal composition.

It may be worth mentioning that when I take any small bundle of portraits, selected at hazard, I have generally found it easy to sort them into about five groups, four of which have enough resemblance among themselves to make as many fairly clear composites, while the fifth consists of faces that are too incongruous to be grouped in a single class. In dealing with portraits of brothers and sisters, I can generally throw most of them into a single group, with success.

In the small collection of composites given in the Plate facing p. 8, I have purposely selected many of those that I have previously published, and whose originals, on a larger scale, I have at various times exhibited, together with their components, in order to put the genuineness of the results beyond doubt. Those who see them for the first time can hardly believe but that one dominant face has overpowered the rest, and that they are composites only in name. When, however, the details are examined, this objection disappears. It is true that with careless photography one face may be allowed to dominate, but with the care that ought to be taken, and with the precautions described in the Appendix, that does not occur. I have often been amused when showing composites and their components to friends, to hear a strong expression of opinion that the predominance of one face was evident, and then on asking which face it was, to discover that they disagreed. I have even known a composite in which one portrait seemed unduly to prevail, to be remade without the component in question, and the result to be much the same as before, showing that the reason of the resemblance was that the rejected portrait had a close approximation to the ideal average picture of the rest.

These small composites give a better notion of the utmost capacity of the process than the larger ones, from which they are reduced.   In the latter, the ghosts of individual peculiarities are more visible, and usually the equal traces left by every member of a moderately-sized group can be made out by careful inspection; but it is hardly possible to do this in the pictures in the Plate, except in a good light and in a very few of the cases.   On the other hand, the larger pictures do not contain more detail of value than the smaller ones.

DESCRIPTION OF THE COMPOSITES.

The medallion of Alexander the Great was made by combining the images of six different medals, with a view of obtaining the type of features that the makers of those medals concurred in desiring to ascribe to him.   The originals were kindly selected for me by Mr. R. Stuart Poole from the collection in the British Museum.   This composite was one of the first I ever made, and is printed together with its six components in the *Journal of the Royal Institution*, in illustration of a lecture I gave there in April 1879.   It seems to me that it is possible on this principle to obtain a truer likeness of a man than in any other way.   Every artist makes mistakes; but by combining the conscientious works of many artists, their separate mistakes disappear, and what is common to all of their works remains.   So as regards different photographs of the same person, those accidental momentary expressions are got rid of, which an ordinary photograph made by a brief exposure cannot help recording.   On the other hand, any happy sudden trait of expression is lost.   The composite gives the features in repose.

The next pair of composites (full face and profile) on the Plate has not been published before.   The interest of the pair lies chiefly in their having been made from only two components, and they show how curiously even two

SPECIMENS OF COMPOSITE PORTRAITURE

PERSONAL AND FAMILY.

Alexander the Great From 6 Different Medals.

Two Sisters.

From 6 Members of same Family Male & Female.

HEALTH. DISEASE. CRIMINALITY.

23 Cases Royal Engineers. 12 Officers. 11 Privates.

 6 Cases

 9 Cases

Tubercular Disease.

 8 Cases

 4 Cases

2 Of the many Criminal Types.

CONSUMPTION AND OTHER MALADIES

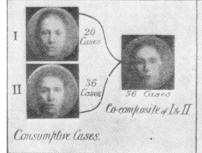

I *20 Cases*

II *36 Cases*

56 Cases

Co-composite of I & II.

Consumptive Cases.

100 Cases

50 Cases

Not Consumptive.

To face page 8

faces that have a moderate family likeness will blend into a single one. That neither of these predominated in the present case will be learned from the following letter by the father of the ladies, who is himself a photographer :—

" I am exceedingly obliged for the very curious and interesting composite portraits of my two children. Knowing the faces so well, it caused me quite a surprise when I opened your letter. I put one of the full faces on the table for the mother to pick up casually. She said, 'When did you do this portrait of A? how *like* she is to B! Or *is* it B? I never thought they were so like before.' It has puzzled several people to say whether the profile was intended for A or B. Then I tried one of them on a friend who has not seen the girls for years. He said, 'Well, it is one of the family for certain, but I don't know which.'"

I have made several other family portraits, which to my eye seem great successes, but must candidly own that the persons whose portraits are blended together seldom seem to care much for the result, except as a curiosity. We are all inclined to assert our individuality, and to stand on our own basis, and to object to being mixed up indiscriminately with others. The same feeling finds expression when the resident in a suburban street insists on calling his house a villa with some fantastic name, and refuses, so long as he can, to call it simply Number so and so in the street.

The last picture in the upper row shows the easy way in which young and old, male and female, combine to form an effective picture. The components consist in this case of the father and mother, two sons, and two daughters. I exhibited the original of this, together with the portraits from which it was taken, at the Loan Photographic Exhibition at the Society of Arts in February 1882. I also sent copies of the original of this same composite to several amateur photographers, with a circular letter asking them to get from me family groups for the purpose of experiments, to see how far the process was suitable for family portraiture.

The middle row of portraits illustrates health, disease, and criminality. For health, I have combined the portraits of twelve officers of the Royal Engineers with about an equal number of privates, which were taken for me by

Lieutenant Darwin, R.E.   The individuals from whom this composite was made, which has not come out as clearly as I should have liked, differed considerably in feature, and they came from various parts of England.   The points they had in common were the bodily and mental qualifications required for admission into their select corps, and their generally British descent.   The result is a composite having an expression of considerable vigour, resolution, intelligence, and frankness.   I have exhibited both this and others that were made respectively from the officers, from the whole collection of privates—thirty-six in number—and from that selected portion of them that is utilised in the present instance.

This face and the qualities it connotes probably gives a clue to the direction in which the stock of the English race might most easily be improved.   It is the essential notion of a race that there should be some ideal typical form from which the individuals may deviate in all directions, but about which they chiefly cluster, and towards which their descendants will continue to cluster.   The easiest direction in which a race can be improved is towards that central type, because nothing new has to be sought out.   It is only necessary to encourage as far as practicable the breed of those who conform most nearly to the central type, and to restrain as far as may be the breed of those who deviate widely from it.   Now there can hardly be a more appropriate method of discovering the central physiognomical type of any race or group than that of composite portraiture.

As a contrast to the composite of the Royal Engineers, I give those of two of the coarse and low types of face found among the criminal classes.   The photographs from which they were made are taken from two large groups. One are those of men undergoing severe sentences for murder and other crimes connected with violence ; the other of thieves.   They were reprints from those taken by order of the prison authorities for purposes of identification. I was allowed to obtain copies for use in my inquiries by the kind permission of Sir Edmund Du Cane, H.M. Director of Prisons.   The originals of these and their components have frequently been exhibited.   It is unhappily a fact that fairly distinct types of criminals breeding true to their kind have become established, and are one of the saddest disfigurements of modern civilisation. To this subject I shall recur.

I have made numerous composites of various groups of convicts, which are interesting negatively rather than positively. They produce faces of a mean description, with no villainy written on them. The individual faces are villainous enough, but they are villainous in different ways, and when they are combined, the individual peculiarities disappear, and the common humanity of a low type is all that is left.

The remaining portraits are illustrations of the application of the process to the study of the physiognomy of disease. They were published a year ago with many others, together with several of the portraits from which they were derived, in a joint memoir by Dr. Mahomed and myself, in vol. xxv. of the *Guy's Hospital Reports*. The originals and all the components have been exhibited on several occasions.

In the lower division of the Plate will be found three composites, each made from a large number of faces, unselected, except on the ground of the disease under which they were suffering. When only few portraits are used, there must be some moderate resemblance between them, or the result would be blurred ; but here, dealing with as many as 56, 100, and 50 cases respectively, the combination of any medley group results in an ideal expression.

It will be observed that the composite of 56 female faces is made by the blending of two other composites, both of which are given. The history was this—I took the 56 portraits and sorted them into two groups ; in the first of these were 20 portraits that showed a tendency to thin features, in the other group there were 36 that showed a tendency to thickened features. I made composites of each of them as shown in the Plate. Now it will be remarked that, notwithstanding the attempt to make two contrasted groups, the number of mediocre cases was so great that the composites of the two groups are much alike. If I had divided the 56 into two haphazard groups, the results would have been closely alike, as I know from abundant experience of the kind. The co-composite of the two will be observed to have an intermediate expression. The test and measure of statistical truth lies in the degree of accordance between results obtained from different batches of instances of the same generic class. It will be gathered from these instances that composite portraiture may attain statistical constancy, within limits not easily distinguished by the eye, after some 30 haphazard portraits of the same class have been combined. This at least has been my experience thus far.

The two faces illustrative of the same type of tubercular disease are very striking ; the uppermost is photographically interesting as a case of predominance of one peculiarity, happily of no harm to the effect of the ideal wan face. It is that one of the patients had a sharply-checked black and white scarf, whose pattern has asserted itself unduly in the composite. In such cases I ought to throw the too clearly defined picture a little out of focus. The way in which the varying brightness of different pictures is reduced to a uniform standard of illumination is described in the Appendix.

It must be clearly understood that these portraits do not profess to give the whole story of the physiognomy of phthisis. I have not room to give illustrations of other types—namely, that with coarse and blunted features, or the strumous one, nor any of the intermediates. These have been discussed chiefly by Dr. Mahomed in the memoir alluded to above.

In the large experience I have had of sorting photographs, literally by the thousand, while making experiments with composites, I have been struck by certain general impressions. The consumptive patients consisted of many hundred cases, including a considerable proportion of very ignoble specimens of humanity. Some were scrofulous and misshapen, or suffered from various loathsome forms of inherited disease ; most were ill nourished. Nevertheless, in studying their portraits the pathetic interest prevailed, and I returned day after day to my tedious work of classification, with a liking for my materials. It was quite otherwise with the criminals. I did not adequately appreciate the degradation of their expressions for some time ; at last the sense of it took firm hold of me, and I cannot now handle the portraits without overcoming by an effort the aversion they suggest.

I am sure that the method of composite portraiture opens a fertile field of research to ethnologists, but I find it very difficult to do much single-handed, on account of the difficulty of obtaining the necessary materials. As a rule, the individuals must be specially photographed. The portraits made by artists are taken in every conceivable aspect and variety of light and shade, but for the purpose in question the aspect and the shade must be the same throughout. Group portraits would do to work from, were it not for the strong out-of-door light under which they are necessarily taken, which gives an unwonted effect to the expression of the faces. Their scale also is too small to give a sufficiently clear picture when enlarged. I may say that the scale of the portraits need not be uniform, as my apparatus enlarges or reduces as required, at the same time that it superposes the images ; but the portraits of the heads should never be less than twice the size of that of the Queen on a halfpenny piece.

I heartily wish that amateur photographers would seriously take up the subject of composite portraiture as applied to different sub-types of the varying races of men. I have already given more time to perfecting the process and experimenting with it than I can well spare.

BODILY QUALITIES.

The differences in the bodily qualities that are the usual subjects of anthropometry are easily dealt with, and are becoming widely registered in many countries. We are unfortunately destitute of trustworthy measurements of Englishmen of past generations to enable us to compare class with class, and to learn how far the several sections of the English nation may be improving or deteriorating. We shall, however, hand useful information concerning our own times to our successors, thanks principally to the exertions of an Anthropometric Committee established five years ago by the British Association, who have collected and partly classified and published a large amount of facts, besides having induced several institutions, such as Marlborough College, to undertake a regular system of anthropometric record. I am not, however, concerned here with the labours of this committee, nor with the separate valuable publications of some of its members, otherwise than in one small particular which appears to show that the English population as a whole, or perhaps I should say the urban portion of it, is in some sense deteriorating. It is that the average stature of the older persons measured by or for the committee has not been found to decrease steadily with their age, but sometimes the reverse.[1] This contradicts observations made on the heights of the same men at different periods, whose stature after middle age is invariably reduced by the shrinking of the cartilages. The explanation offered was that the statistical increase of stature with age should be ascribed to the survival of the more stalwart. On reconsideration, I am inclined to doubt the adequacy of the explanation, and partly to account for the fact by a steady, slight deterioration of stature in successive years ; in the urban population owing to the conditions of their lives, and in the rural population owing to the continual draining away of the more stalwart of them to the towns.

It cannot be doubted that town life is harmful to the town population. I have myself investigated its effect on fertility (see Appendix B), and found that taking on the one hand a number of rural parishes, and on the other hand the inhabitants of a medium town, the former reared nearly twice as many adult grandchildren as the latter. The vital functions are so closely related that an inferiority in the production of healthy children very probably implies a loss of vigour generally, one sign of which is a diminution of stature.

Though the bulk of the population may deteriorate, there are many signs that the better housed and fed portion of it improves. In the earlier years of this century the so-called manly sports of boxing and other feats of strength ranked high among the national amusements. A man who was

[1] *Trans. Brit. Assoc.*, 1881, Table V., p. 242 ; and remarks by Mr. Roberts, p. 235.

successful in these became the hero of a large and demonstrative circle of admirers, and it is to be presumed that the best boxer, the best pedestrian, and so forth, was the best adapted to succeed, through his natural physical gifts. If he was not the most gifted man in those respects in the whole kingdom, he was certainly one of the most gifted of them. It therefore does no injustice to the men of that generation to compare the feats of their foremost athletes with those of ours who occupy themselves in the same way. The comparison would probably err in their favour, because the interest in the particular feats in which our grandfathers and great-grandfathers delighted are not those that chiefly interest the present generation, and notwithstanding our increased population, there are fewer men now who attempt them. In the beginning of this century there were many famous walking matches, and incomparably the best walker was Captain Barclay of Ury. His paramount feat, which was once very familiar to the elderly men of the present time, was that of walking a thousand miles in a thousand hours, but of late years that feat has been frequently equalled and overpassed. I am willing to allow much influence to the modern conditions of walking under shelter and subject to improved methods of training (Captain Barclay himself originated the first method, which has been greatly improved since his time); still the fact remains that in executing this particular feat, the athletes of the present day are more successful than those who lived some eighty years ago.

I may be permitted to give an example bearing on the increased stature of the better housed and fed portion of the nation, in a recollection of my own as to the difference in height between myself and my fellow-collegians at Trinity College, Cambridge, in 1840–4. My height is 5 feet $9\frac{3}{4}$ inches, and I recollect perfectly that among the crowd of undergraduates I stood somewhat taller than the majority. I generally looked a little downward when I met their eyes. In later years, whenever I have visited Cambridge, I have lingered in the ante-chapel and repeated the comparison, and now I find myself decidedly shorter than the average of the students. I have precisely the same kind of recollection and the same present experience of the height of crowds of well-dressed persons. I used always to get a fair view of

what was going on over or between their heads; I rarely
can do so now.

The athletic achievements at school and college are much
superior to what they used to be.   Part is no doubt due to
more skilful methods of execution, but not all.   I cannot
doubt that the more wholesome and abundant food, the
moderation in drink, the better cooking, the warmer wear-
ing apparel, the airier sleeping rooms, the greater cleanliness,
the more complete change in holidays, and the healthier
lives led by the women in their girlhood, who become
mothers afterwards, have a great influence for good on the
favoured portion of our race.

The proportion of weakly and misshapen individuals is
not to be estimated by those whom we meet in the streets;
the worst cases are out of sight.   We should parade before
our mind's eye the inmates of the lunatic, idiot, and pauper
asylums, the prisoners, the patients in hospitals, the sufferers
at home, the crippled, and the congenitally blind, and that
large class of more or less wealthy persons who flee to the
sunnier coasts of England, or expatriate themselves for the
chance of life.   There can hardly be a sadder sight than
the crowd of delicate English men and women with narrow
chests and weak chins, scrofulous, and otherwise gravely
affected, who are to be found in some of these places.
Even this does not tell the whole of the story; if there
were a conscription in England, we should find, as in other
countries, that a large fraction of the men who earn their
living by sedentary occupations are unfit for military service.
Our human civilised stock is far more weakly through congenital imperfection than that of any other species of animals,
whether wild or domestic.

It is, however, by no means the most shapely or the
biggest personages who endure hardship the best.   Some
very shabby-looking men have extraordinary stamina.
Sickly-looking and puny residents in towns may have a
more suitable constitution for the special conditions of
their lives, and may in some sense be better knit and do
more work and live longer than much haler men imported
to the same locality from elsewhere.   A wheel and a barrel
seem to have the flimsiest possible constitutions; they

consist of numerous separate pieces all oddly shaped, which, when lying in a heap, look hopelessly unfitted for union; but put them properly together, compress them with a tire in the one case and with hoops in the other, and a remarkably enduring organisation will result. A wheel with a ton weight on the top of it in the waggons of South Africa will jolt for thousands of miles over stony, roadless country without suffering harm; a keg of water may be strapped on the back of a pack-ox or a mule, and be kicked off and trampled on, and be otherwise misused for years, without giving way.

I do not propose to enter further into the anthropometric differences of race, for the subject is a very large one, and this book does not profess to go into detail. Its intention is to touch on various topics more or less connected with that of the cultivation of race, or, as we might call it, with "eugenic"[1] questions, and to present the results of several of my own separate investigations.

ENERGY.

Energy is the capacity for labour. It is consistent with all the robust virtues, and makes a large practice of them possible. It is the measure of fulness of life; the more energy the more abundance of it; no energy at all is death; idiots are feeble and listless. In the inquiries I made on the antecedents of men of science no points came out more strongly than that the leaders of scientific thought were generally gifted with remarkable energy, and that they had

[1] That is, with questions bearing on what is termed in Greek, *eugenes*, namely, good in stock, hereditarily endowed with noble qualities. This, and the allied words, *eugeneia*, etc., are equally applicable to men, brutes, and plants. We greatly want a brief word to express the science of improving stock, which is by no means confined to questions of judicious mating, but which, especially in the case of man, takes cognisance of all influences that tend in however remote a degree to give to the more suitable races or strains of blood a better chance of prevailing speedily over the less suitable than they otherwise would have had. The word *eugenics* would sufficiently express the idea; it is at least a neater word and a more generalised one than *viriculture*, which I once ventured to use.

C

inherited the gift of it from their parents and grandparents. I have since found the same to be the case in other careers.

Energy is an attribute of the higher races, being favoured beyond all other qualities by natural selection. We are goaded into activity by the conditions and struggles of life. They afford stimuli that oppress and worry the weakly, who complain and bewail, and it may be succumb to them, but which the energetic man welcomes with a good-humoured shrug, and is the better for in the end.

The stimuli may be of any description : the only important matter is that all the faculties should be kept working to prevent their perishing by disuse. If the faculties are few, very simple stimuli will suffice. Even that of fleas will go a long way. A dog is continually scratching himself, and a bird pluming itself, whenever they are not occupied with food, hunting, fighting, or love. In those blank times there is very little for them to attend to besides their varied cutaneous irritations. It is a matter of observation that well washed and combed domestic pets grow dull; they miss the stimulus of fleas. If animals did not prosper through the agency of their insect plagues, it seems probable that their races would long since have been so modified that their bodies should have ceased to afford a pasture-ground for parasites.

It does not seem to follow that because men are capable of doing hard work they like it. Some, indeed, fidget and fret if they cannot otherwise work off their superfluous steam ; but on the other hand there are many big lazy fellows who will not get up their steam to full pressure except under compulsion. Again, the character of the stimulus that induces hard work differs greatly in different persons ; it may be wealth, ambition, or other object of passion. The solitary hard workers, under no encouragement or compulsion except their sense of duty to their generation, are unfortunately still rare among us.

It may be objected that if the race were too healthy and energetic there would be insufficient call for the exercise of the pitying and self-denying virtues, and the character of men would grow harder in consequence. But it does not seem reasonable to preserve sickly breeds for the sole purpose of tending them, as the breed of foxes is preserved solely for sport and its attendant advantages. There is little fear that misery will ever cease from the land, or that the compassionate will fail to find objects for their compassion; but at present the supply vastly exceeds the demand: the land is overstocked and overburdened with the listless and the incapable.

In any scheme of eugenics, energy is the most important quality to favour; it is, as we have seen, the basis of living action, and it is eminently transmissible by descent.

SENSITIVITY.

The only information that reaches us concerning outward events appears to pass through the avenue of our senses; and the more perceptive the senses are of difference, the larger is the field upon which our judgment and intelligence can act. Sensation mounts through a series of grades of "just perceptible differences." It starts from the zero of consciousness, and it becomes more intense as the stimulus increases (though at a slower rate) up to the point when the stimulus is so strong as to begin to damage the nerve apparatus. It then yields place to pain, which is another form of sensation, and which continues until the nerve apparatus is destroyed. Two persons may be equally able just to hear the same faint sound, and they may equally begin to be pained by the same loud sound, and yet they may differ as to the number of intermediate grades of sensation. The grades will be less numerous as the organisation is of a lower order, and the keenest sensation possible to it will in consequence be less intense. An artist who is capable of discriminating more differences of tint than another man is not necessarily more capable of seeing clearly in twilight, or more or less intolerant of sunshine. A musician is not necessarily able to hear very faint sounds, nor to be more startled by loud sounds than others are. A mechanic who works hard with heavy tools and has rough and grimy thumbs, insensible to very slight pressures, may yet have a singularly discriminating power of touch in respect to the pressures that he can feel.

The discriminative faculty of idiots is curiously low;

they hardly distinguish between heat and cold, and their sense of pain is so obtuse that some of the more idiotic seem hardly to know what it is. In their dull lives, such pain as can be excited in them may literally be accepted with a welcome surprise. During a visit to Earlswood Asylum I saw two boys whose toe-nails had grown into the flesh and had been excised by the surgeon. This is a horrible torture to ordinary persons, but the idiot lads were said to have shown no distress during the operation; it was not necessary to hold them, and they looked rather interested at what was being done.[1] I also saw a boy with the scar of a severe wound on his wrist; the story being that he had first burned himself slightly by accident, and, liking the keenness of the new sensation, he took the next opportunity of repeating the experience, but, idiot-like, he overdid it.

The trials I have as yet made on the sensitivity of different persons confirms the reasonable expectation that it would on the whole be highest among the intellectually ablest. At first, owing to my confusing the quality of which I am speaking with that of nervous irritability, I fancied that women of delicate nerves who are distressed by noise, sunshine, etc., would have acute powers of discrimination. But this I found not to be the case. In morbidly sensitive persons both pain and sensation are induced by lower stimuli than in the healthy, but the number of just perceptible grades of sensation between them is not necessarily different.

I found as a rule that men have more delicate powers of discrimination than women, and the business experience of life seems to confirm this view. The tuners of pianofortes are men, and so I understand are the tasters of tea and wine, the sorters of wool, and the like. These latter occupations are well salaried, because it is of the first moment to the merchant that he should be rightly advised on the real value of what he is about to purchase or to sell. If the sensitivity of women were superior to that of men, the self-interest of merchants would lead to their being

[1] See "Remarks on Idiocy," by E. W. Graham, M.D., *Medical Journal*, January 16, 1875.

always employed; but as the reverse is the case, the opposite supposition is likely to be the true one.

Ladies rarely distinguish the merits of wine at the dinner-table, and though custom allows them to preside at the breakfast-table, men think them on the whole to be far from successful makers of tea and coffee.

Blind persons are reputed to have acquired in compensation for the loss of their eyesight an increased acuteness in their other senses; I was therefore curious to make some trials with my test apparatus, which I will describe in the next chapter. I was permitted to do so on a number of boys at a large educational blind asylum, but found that, although they were anxious to do their best, their performances were by no means superior to those of other boys. It so happened that the blind lads who showed the most delicacy of touch and won the little prizes I offered to excite emulation, barely reached the mediocrity of the various sighted lads of the same age whom I had previously tested. I have made not a few observations and inquiries, and find that the guidance of the blind depends mainly on the multitude of collateral indications to which they give much heed, and not in their superior sensitivity to any one of them. Those who see do not care for so many of these collateral indications, and habitually overlook and neglect several of them. I am convinced also that not a little of the popular belief concerning the sensitivity of the blind is due to exaggerated claims on their part that have not been verified. Two instances of this have fallen within my own experience, in both of which the blind persons claimed to have the power of judging by the echo of their voice and by certain other feelings, the one when they were approaching objects, even though the object was so small as a handrail, and the other to tell how far the door of the room in which he was standing was open. I used all the persuasion I could to induce each of these persons to allow me to put their assertions to the test; but it was of no use. The one made excuses, the other positively refused. They had probably the same tendency that others would have who happened to be defective in any faculty that their comrades possessed, to fight bravely against their disadvantage, and at the same time to be betrayed into some overvaunting of their capacities in other directions. They would be a little conscious of this, and would therefore shrink from being tested.

The power of reading by touch is not so very wonderful. A former Lord Chancellor of England, the late Lord Hatherley, when he was advanced in years, lost his eyesight for some time owing to a cataract, which was not ripe to be operated on. He assured me that he had then learned and practised reading by touch very rapidly. This fact may perhaps also serve as additional evidence of the sensitivity of able men.

Notwithstanding many travellers' tales, I have thus far been unsuccessful in obtaining satisfactory evidence of any general large superiority of the senses of savages over those of civilised men. My own experience, so far as it goes, of Hottentots, Damaras, and some other wild races, went to show that their sense discrimination was not superior to those of white men, even as regards keenness of eyesight. An offhand observer is apt to err by assigning to a single cause what is partly due to others as well. Thus, as regards eyesight, a savage who is accustomed to watch oxen grazing at a distance becomes so familiar with their appearance and habits that he can identify particular animals and draw conclusions as to what they are doing with an accuracy that may seem to strangers to be wholly dependent on exceptional acuteness of vision. A sailor has the reputation of keen sight because he sees a distant coast when a lands-man cannot make it out; the fact being that the landsman usually expects a different appearance to what he has really to look for, such as a very minute and sharp outline instead of a large, faint blur. In a short time a landsman becomes quite as quick as a sailor, and in some test experiments [1] he was found on the average to be distinctly the superior. It is not surprising that this should be so, as sailors in vessels of moderate size have hardly ever the practice of focussing their eyes sharply upon objects farther off than the length of the vessel or the top of the mast, say at a distance of fifty paces. The horizon itself as seen from the deck, and

[1] Gould's *Military and Anthropological Statistics*, p. 530. New York, 1869.

under the most favourable circumstances, is barely four miles off, and there is no sharpness of outline in the intervening waves. Besides this, the life of a sailor is very unhealthy, as shown by his growing old prematurely, and his eyes must be much tried by foul weather and salt spray.

We inherit our language from barbarous ancestors, and it shows traces of its origin in the imperfect ways by which grades of difference admit of being expressed. Suppose a pedestrian is asked whether the knapsack on his back feels heavy. He cannot find a reply in two words that cover more varieties than (1) very heavy, (2) rather heavy, (3) moderate, (4) rather light, (5) very light. I once took considerable pains in the attempt to draw up verbal scales of more than five orders of magnitude, using those expressions only that every cultivated person would understand in the same sense; but I did not succeed. A series that satisfied one person was not interpreted in the same sense by another.

The general intention of this chapter has been to show that a delicate power of sense discrimination is an attribute of a high race, and that it has not the drawback of being necessarily associated with nervous irritability.

Sequence of Test Weights.

I will now describe an apparatus I have constructed to test the delicacy with which weights may be discriminated by handling them. I do so because the principle on which it is based may be adopted in apparatus for testing other senses, and its description and the conditions of its use will illustrate the desiderata and difficulties of all such investigations.

A series of test weights is a simple enough idea —the difficulty lies in determining the particular sequence of weights that should be employed. Mine form a geometric series, for the reason that when stimuli of all kinds increase by geometric grades the sensations they give rise to will increase by arithmetic grades, so long as the stimulus is neither so weak as to be barely felt, nor so strong as to excite fatigue. My apparatus, which is explained more fully in the Appendix, consists of a number of common gun cartridge cases filled with alternate layers of shot, wool, and wadding, and then closed in the usual way. They are all identical in appearance, and may be said to differ only in their specific gravities. They are marked in numerical sequence with the register numbers, 1, 2, 3, etc., but their weights are proportioned to the numbers of which 1, 2, 3, etc., are the logarithms, and consequently run in a geometric series. Hence the numbers of the weights form a scale of equal degrees of sensitivity. If a person can just distinguish between the weights numbered 1 and 3, he can also just distinguish between 2 and 4, 3 and 5, and any other pair of weights of which the register number of the one exceeds that of the other by 2. Again, his coarseness of discrimination is exactly double of that of another person who can just distinguish pairs of weights differing only by 1, such as 1 and 2, 2 and 3, 3 and 4, and so on. The testing is performed by handing pairs of weights to the operatee until his power of discrimination is approximately made out, and then to proceed more carefully. It is best now, for reasons stated in the Appendix, to hand to the operatee sequences of three weights at a time, after shuffling them. These he has to arrange in their proper order, with his eyes shut, and by the sense of their weight alone. The operator finally records the scale interval that the operatee can just appreciate, as being the true measure of the coarseness (or the inverse measure of the delicacy) of the sensitivity of the operatee.

It is somewhat tedious to test many persons in succession, but any one can test his own powers at odd and end times with ease and nicety, if he happens to have ready access to suitable apparatus.

The use of tests, which, objectively speaking, run in a geometric series, and subjectively in an arithmetic one, may be applied to touch, by the use of wire-work of various degrees of fineness ; to taste, by stock bottles of solutions of salt, etc., of various strengths ; to smell, by bottles of attar of rose, etc., in various degrees of dilution.

The tests show the sensitivity at the time they are made, and give an approximate measure of the discrimination with which the operatee habitually employs his senses. It does not measure his capacity for discrimination, because the discriminative faculty admits of much education, and the test results always show increased delicacy after a little practice. However, the requirements of everyday life educate all our faculties in some degree, and I have not found the performances with test weights to improve much after a little familiarity with their use. The weights have, as it were, to be played with at first, then they must be tried carefully on three or four separate occasions.

I did not at first find it at all an easy matter to make test weights so alike as to differ in no other appreciable respect than in their specific gravity, and if they differ and become known apart, the knowledge so acquired will vitiate future judgments in various indirect ways. Similarity in outward shape and touch was ensured by the use of mechanically-made cartridge cases; dissimilarity through any external stain was rendered of no hindrance to the experiment by making the operatee handle them in a bag or with his eyes shut. Two bodies may, however, be alike in weight and outward appearance and yet behave differently when otherwise mechanically tested, and, consequently, when they are handled. For example, take two eggs, one raw and the other hard boiled, and spin them on the table; press the finger for a moment upon either of them whilst it is still spinning: if it be the hard-boiled egg it will stop as dead as a stone: if it be the raw egg, after a little apparent hesitation, it will begin again to rotate. The motion of its shell had alone been stopped; the internal part was still rotating and this compelled the shell to follow it. Owing to this cause, when we handle the two eggs, the one feels "quick" and the other does not. Similarly with the cartridges, when one is rather more loosely packed than the others the difference is perceived on handling them. Or it may have one end heavier than the other, or else its weight may not be equally distributed round its axis, causing it to rest on the table with the same part always lowermost; differences due to these causes are also easily perceived when handling the cartridges. Again, one of two similar cartridges may balance perfectly in all directions, but the weight of one of them may be disposed too much towards the ends, as in a dumb-bell, or gathered too much towards the centre. The period of oscillation will differ widely in the two cases, as may be shown by suspending the cartridges by strings round their middle so that they shall hang horizontally, and then by a slight tap making them spin to and fro round the string as an axis.

The touch is very keen in distinguishing all these peculiarities. I have mentioned them, and might have added more, to show that experiments on sensitivity have to be made in the midst of pitfalls warily to be avoided. Our apparently simplest perceptions are very complex. We hardly ever act on the information given by only one element of one sense, and our sensitivity in any desired direction cannot be rightly determined except by carefully-devised apparatus judiciously used.

WHISTLES FOR AUDIBILITY OF SHRILL NOTES.

I contrived a small whistle for conveniently ascertaining the upper limits of audible sound in different persons, which Dr. Wollaston had shown to vary considerably. He used small pipes, and found much difficulty in making them. I made a very small whistle from a brass tube whose internal diameter was less than one-tenth of an inch in diameter. A plug was fitted into the lower end of the tube, which could be pulled out or pushed in as much as desired, thereby causing the length of the bore of the whistle to be varied at will. When the bore is long the note is low ; when short, it is high. The plug was graduated, so that the precise note produced by the whistle could be determined by reading off the graduations and referring to a table. (See Appendix.)

On testing different persons, I found there was a remarkable falling off in the power of hearing high notes as age advanced. The persons themselves were quite unconscious of their deficiency so long as their sense of hearing low notes remained unimpaired. It is an only too amusing experiment to test a party of persons of various ages, including some rather elderly and self-satisfied personages. They are indignant at being thought deficient in the power of hearing, yet the experiment quickly shows that they are absolutely deaf to shrill notes which the younger persons hear acutely, and they commonly betray much dislike to the discovery. Every one has his limit, and the limit at which sounds become too shrill to be audible to any particular person can be rapidly determined by this little instrument. Lord Rayleigh and others have found that sensitive flames are powerfully affected by the vibrations of whistles that are too rapid to be audible to ordinary ears.

I have tried experiments with all kinds of animals on their powers of hearing shrill notes. I have gone through the whole of the Zoological Gardens, using an apparatus arranged for the purpose. It consists of one of my little whistles at the end of a walking-stick—that is, in reality, a long tube ; it has a bit of india-rubber pipe under the handle, a sudden squeeze upon which forces a little air into the whistle and causes it to sound. I hold it as near as is safe to the ears of the animals, and when they are quite accustomed to its presence and heedless of it, I make it sound ; then if they prick their ears it shows that they hear the whistle ; if they do not, it is probably inaudible to them. Still, it is very possible that in some cases they hear but do not heed the sound. Of all creatures, I have found none superior to cats in the power of hearing shrill sounds ; it is perfectly remarkable what a faculty they have in this way. Cats, of course, have to deal in the dark with mice, and to find them out by their squealing. Many people cannot hear the shrill squeal of a mouse. Some time ago, singing mice were exhibited in London, and of the people who went to hear them, some could hear nothing, whilst others could hear a little, and others again could hear much. Cats are differentiated by natural selection until they have a power of hearing all the high notes made by mice and other little creatures that they have to catch. A cat that is at a very considerable distance, can be made to turn its ear round by sounding a note that is too shrill to be audible by almost any human ear. Small dogs also hear very shrill notes, but large ones do not. I have walked through the streets of a town with an instrument like that which I used in the Zoological Gardens, and made nearly all the little dogs turn round, but not the large ones. At Berne, where there appear to be more large dogs lying idly about the streets than in any other town in Europe, I have tried the whistle for hours together, on a great many large dogs, but could not find one that heard it. Ponies are sometimes able to hear very high notes. I once frightened a pony with one of these whistles in the middle of a large field. My attempts on insect hearing have been failures.

ANTHROPOMETRIC REGISTERS.

When shall we have anthropometric laboratories, where a man may, when he pleases, get himself and his children weighed, measured, and rightly photographed, and have their bodily faculties tested by the best methods known to modern science? The records of growth of numerous persons from childhood to age are required before it can be possible to rightly appraise the effect of external conditions upon development, and records of this kind are at present non-existent. The various measurements should be accompanied by photographic studies of the features in full face and in profile, and on the same scale, for convenience of comparison.

We are all lazy in recording facts bearing on ourselves, but parents are glad enough to do so in respect to their children, and they would probably be inclined to avail themselves of a laboratory where all that is required could be done easily and at small cost. These domestic records would hereafter become of considerable biographical interest. Every one of us in his mature age would be glad of a series of pictures of himself from childhood onwards, accompanied by physical records, and arranged consecutively with notes of current events by their sides. Much more would he be glad of similar collections referring to his father, mother, grandparents, and other near relatives. It would be peculiarly grateful to the young to possess likenesses of their parents and those whom they look upon as heroes, taken when they were of the same age as themselves. Boys are too apt to think of their parents as having always been elderly men, because they have insufficient data to construct imaginary pictures of them as they were in their youth.

The cost of taking photographs in batches is so small, and

the time occupied is so brief, when the necessary preparations have been made and the sitters are ready at hand, that a practice of methodically photographing schoolboys and members of other large institutions might easily be established. I, for one, should dearly prize the opportunity of visiting the places where I have been educated, and of turning over pages showing myself and my companions as we were in those days. But no such records exist ; the institutions last and flourish, the individuals who pass through them are dispersed and leave few or no memorials behind. It seems a cruel waste of opportunity not to make and keep these brief personal records in a methodical manner. The fading of ordinary photographic prints is no real objection to keeping a register, because they can now be reproduced at small charge in permanent printers' ink, by the autotype and other processes.

I have seen with admiration, and have had an opportunity of availing myself of, the newly-established library of well-ordered folios at the Admiralty, each containing a thousand pages, and each page containing a brief summary of references to the life of a particular seaman. There are already 80,000 pages, and owing to the excellent organisation of the office it is a matter of perfect ease to follow out any one of these references, and to learn every detail of the service of any seaman. A brief register of measurements and events in the histories of a large number of persons, previous to their entering any institution and during their residence in it, need not therefore be a difficult matter to those who may take it in hand seriously and methodically.

The recommendation I would venture to make to my readers is to obtain photographs and ordinary measurements periodically of themselves and their children, making it a family custom to do so, because, unless driven by some custom, the act will be postponed until the opportunity is lost. Let those periodical photographs be full and side views of the face on an adequate scale, adding any others that may be wished, but not omitting these. As the portraits accumulate have collections of them autotyped. Keep the prints methodically in a family register, writing by their side careful chronicles of illness and all such events as used to find a place on the fly-leaf of the Bible of former generations, and inserting other interesting personal facts and whatever anthropometric data can be collected.

Those who care to initiate and carry on a family chronicle illustrated by abundant photographic portraiture, will produce a work that they and their children and their descendants in more remote generations will assuredly be grateful for. The family tie has a real as well as a traditional significance. The world is beginning to awaken to the fact that the life of the individual is in some real sense a prolongation of those of his ancestry. His vigour, his character, and his diseases are principally derived from theirs; sometimes his faculties are blends of ancestral qualities; but more frequently they are mosaics, patches of resemblance to one or other of them showing now here and now there. The life-histories of our relatives are prophetic of our own futures; they are far more instructive to us than those of strangers, far more fitted to encourage and to forewarn us. If there be such a thing as a natural birthright, I can conceive of none superior to the right of the child to be informed, at first by proxy through his guardians, and afterwards personally, of the life-history, medical and other, of his ancestry. The child is thrust into existence without his having any voice at all in the matter, and the smallest amend that those who brought him here can make, is to furnish him with all the guidance they can, including the complete life-histories of his near progenitors.

The investigation of human eugenics—that is, of the conditions under which men of a high type are produced—is at present extremely hampered by the want of full family histories, both medical and general, extending over three or four generations. There is no such difficulty in investigating animal eugenics, because the generations of horses, cattle, dogs, etc., are brief, and the breeder of any such stock lives long enough to acquire a large amount of experience from his own personal observation. A man, however, can rarely be familiar with more than two or three generations of his contemporaries before age has begun to check his powers; his working experience must therefore be chiefly based upon records. Believing, as I do, that human eugenics will become recognised before long as a study of the highest practical importance, it seems to me that no time ought to be lost in encouraging and directing a habit of compiling personal and family histories.   If the necessary materials be brought into existence, it will require no more than zeal and persuasiveness on the part of the future investigator to collect as large a store of them as he may require.

UNCONSCIOUSNESS OF PECULIARITIES.

The importance of submitting our faculties to measurement lies in the curious unconsciousness in which we are apt to live of our personal peculiarities, and which our intimate friends often fail to remark.   I have spoken of the ignorance of elderly persons of their deafness to high notes, but even the existence of such a peculiarity as colour blindness was not suspected until the memoir of Dalton in 1794. That one person out of twenty-nine or thereabouts should be unable to distinguish a red from a green, without knowing that he had any deficiency of colour sense, and without betraying his deficiency to his friends, seems perfectly incredible to the other twenty-eight; yet as a matter of fact he rarely does either the one or the other.   It is hard to convince the colour-blind of their own infirmity.   I have seen curious instances of this : one was that of a person by no means unpractised in physical research, who had been himself tested in matching colours.   He gave me his own version of the result, to the effect that though he might perhaps have fallen a little short of perfection as judged by over-refined tests, his colour sense was for all practical purposes quite good.   On the other hand, the operator assured me that when he had toned the intensities of a pure red and a pure green in a certain proportion, the person ceased to be able to distinguish between them !   Colour blindness is often very difficult to detect, because the test hues and tints may be discriminated by other means than by the normal colour sense.   Ordinary pigments are never pure, and the test colours may be distinguished by those of their adventitious hues to which the partly colour-blind man may be sensitive.   We do not suspect ourselves to be yellow-blind by candle light, because we enjoy pictures in the evening nearly or perhaps quite as much as in the day time ; yet we may observe that a yellow primrose laid on the white table-cloth wholly loses its colour by candle light, and becomes as white as a snowdrop.

In the inquiries I made on the hereditary transmission of capacity, I was often amused by the naïve remark of men who had easily distanced their competitors, that they ascribed their success to their own exertions. They little recognised how much they owed to their natural gifts of exceptional capacity and energy on the one hand, and of exceptional love for their special work on the other.

In future chapters I shall give accounts of persons who have unusual mental characteristics as regards imagery, visualised numerals, colours connected with sounds and special associations of ideas, being unconscious of their peculiarities; but I cannot anticipate these subjects here, as they all require explanation. It will be seen in the end how greatly metaphysicians and psychologists may err, who assume their own mental operations, instincts, and axioms to be identical with those of the rest of mankind, instead of being special to themselves. The differences between men are profound, and we can only be saved from living in blind unconsciousness of our own mental peculiarities by the habit of informing ourselves as well as we can of those of others. Examples of the success with which this can be done will be found farther on in the book.

I may take this opportunity of remarking on the well-known hereditary character of colour blindness in connection with the fact, that it is nearly twice as prevalent among the Quakers as among the rest of the community, the proportions being as 5·9 to 3·5 per cent.[1] We might have expected an even larger ratio. Nearly every Quaker is descended on both sides solely from members of a group of men and women who segregated themselves from the rest of the world five or six generations ago; one of their strongest opinions being that the fine arts were worldly snares, and their most conspicuous practice being to dress in drabs. A born artist could never have consented to separate himself from his fellows on such grounds; he would have felt the profession of those opinions and their accompanying

[1] *Trans. Ophthalmological Soc.*, 1881, p. 198.

practices to be a treason to his æsthetic nature. Consequently few of the original stock of Quakers are likely to have had the temperament that is associated with a love for colour, and it is in consequence most reasonable to believe that a larger proportion of colour-blind men would have been found among them than among the rest of the population.

Again, Quakerism is a decreasing sect, weakened by yearly desertions and losses, especially as the act of marriage with a person who is not a member of the Society is necessarily followed by exclusion from it. It is most probable that a large proportion of the deserters would be those who, through reversion to some bygone ancestor, had sufficient artistic taste to make a continuance of Quaker practices too irksome to be endured. Hence the existing members of the Society of Friends are a race who probably contained in the first instance an unduly large proportion of colour-blind men, and from whose descendants many of those who were not born colour blind have year by year been drafted away. Both causes must have combined with the already well-known tendency of colour blindness to hereditary transmission, to cause it to become a characteristic of their race. Dalton, who first discovered its existence, as a personal peculiarity of his own, was a Quaker to his death; Young, the discoverer of the undulatory theory of light, and who wrote specially on colours, was a Quaker by birth, but he married outside the body and so ceased to belong to it.

STATISTICAL METHODS.

The object of statistical science is to discover methods of condensing information concerning large groups of allied facts into brief and compendious expressions suitable for discussion. The possibility of doing this is based on the constancy and continuity with which objects of the same species are found to vary. That is to say, we always find, after sorting any large number of such objects in the order (let us suppose) of their lengths, beginning with the shortest and ending with the tallest, and setting them side by side like a long row of park palings between the same limits, their upper outline will be identical. Moreover, it will run smoothly and not in irregular steps. The theoretical interpretation of the smoothness of outline is that the individual differences in the objects are caused by different combinations of a large number of minute influences; and as the difference between any two adjacent objects in a long row must depend on the absence in one of them of some single influence, or of only a few such, that were present in the other, the amount of difference will be insensible. Whenever we find on trial that the outline of the row is not a flowing curve, the presumption is that the objects are not all of the same species, but that part are affected by some large influence from which the others are free; consequently there is a confusion of curves. This presumption is never found to be belied.

It is unfortunate for the peace of mind of the statistician that the influences by which the magnitudes, etc., of the objects are determined can seldom if ever be roundly classed into large and small, without intermediates. He is tantalised by the hope of getting hold of sub-groups of sufficient size that shall contain no individuals except those belonging strictly to the same species, and he is almost constantly baffled. In the end he is obliged to exercise his judgment as to the limit at which he should cease to subdivide. If he subdivides very frequently, the groups become too small to have statistical value; if less frequently, the groups will be less truly specific.

A species may be defined as a group of objects whose individual differences are wholly due to different combinations of the same set of minute causes, no one of which is so powerful as to be able by itself to make any sensible difference in the result. A well-known mathematical consequence flows from this, which is also universally observed as a fact, namely, that in all species the number of individuals who differ from the average value, up to any given amount, is much greater than the number who differ more than that amount, and up to the double of it. In short, if an assorted series be represented by upright lines arranged side by side along a horizontal base at equal distances apart, and of lengths proportionate to the magnitude of the quality in the corresponding objects, then their shape will always resemble that shown in Fig. 1.

The form of the bounding curve resembles that which is called in architectural language an ogive, from "augive," an old French word for a cup, the figure being not unlike the upper half of a cup lying sideways with its axis horizontal. In consequence of the multitude of mediocre values, we always find that on either side of the middlemost ordinate Cc, which is the median value and may be accepted as the average, there is a much less rapid change of height than elsewhere. If the figure were pulled out sideways to make it accord with such physical conceptions as that of a

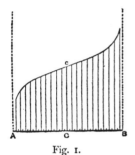

Fig. 1.   Fig. 2.

row of men standing side by side, the middle part of the curve would be apparently horizontal.

The mathematical conception of the curve is best expressed in Fig. 2, where PQ represents any given deviation from the average value, and the ratio of PO to AB represents the relative probability of its occurrence. The equation to the curve and a discussion of its properties will be found in the *Proceedings of the Royal Society*, No. 198, 1879, by Dr. M'Alister. The title of the paper is the "Law of the Geometric Mean," and it follows one by myself on "The Geometric Mean in Vital and Social Statistics."

We can lay down the ogive of any quality, physical or mental, whenever we are capable of judging which of any two members of the group we are engaged upon has the larger amount of that quality. I have called this the method of statistics by intercomparison. There is no bodily or mental attribute in any race of individuals that can be so dealt with, whether our judgment in comparing them be guided by common-sense observation or by actual measurement, which cannot be gripped and consolidated into an ogive with a smooth outline, and thenceforward be treated in discussion as a single object.

It is easy to describe any given ogive which has been based upon measurements, so that it may be drawn from the description with approximate truth. Divide AB into a convenient number of fractional parts, and record the height of the ordinates at those parts. In reproducing the ogive from these data, draw a base line of any convenient length, divide it in the same number of fractional parts, erect ordinates of the stated lengths at those parts, connect their tops with a flowing line, and the thing is done. The most convenient fractional parts are the middle (giving the median), the outside quarters (giving the upper and lower quartiles), and similarly the upper and lower octiles or deciles. This is sufficient for most purposes. It leaves only the outer eighths or tenths of the cases undescribed and undetermined, except so far as may be guessed by the run of the intermediate portion of the curve, and it defines all of the intermediate portion with as close an approximation as is needed for ordinary or statistical purposes.

Thus the heights of all but the outer tenths of the whole body of adult males of the English professional classes may be derived from the five following ordinates, measured in inches, of which the outer pair are deciles :—

$$67\cdot2 ;\ 67\cdot5 ;\ 68\cdot8 ;\ 70\cdot3 ;\ 71\cdot4.$$

Many other instances will be found in the Report of the Anthropometric Committee of the British Association in 1881, pp. 245–257.

When we desire to compare any two large statistical groups, we may compare median with median, quartiles with quartiles, and octiles with octiles ; or we may proceed on the method to be described in the next paragraph but one.

We are often called upon to define the position of an individual in his own series, in which case it is most conformable to usage to give his centesimal grade—that is, his place on the base line AB—supposing it to be graduated from 0° to 100°. In reckoning this, a confusion ought to be avoided between "graduation" and "rank," though it leads to no sensible error in practice. The first of the "park palings" does not stand at A, which is 0°, nor does the hundredth stand at B, which is 100°, for that would make 101 of them ; but they stand at 0°·5 and 99°·5 respectively. Similarly, all intermediate *ranks* stand half a degree short of the *graduation* bearing the same number. When the class is large, the value of half a place becomes extremely small, and the rank and graduation may be treated as identical.

Examples of method of calculating a centesimal position :—

1. A child A is classed after examination as No. 5 in a class of 27 children ; what is his centesimal graduation?

Answer.— If AB be divided into 27 graduations, his rank of No. 5 will correspond to the graduation 4°·5 ; therefore if AB be graduated afresh into 100 graduations, his centesimal grade, x, will be found by the Rule of Three, thus—

$$x : 4°·5 :: 100 : 27 ; \quad x = \tfrac{450°}{27} = 16°·6.$$

2. Another child B is classed No. 13 in a class of 25

Answer.—If AB be divided into 25 graduations, the rank of No. 13 will correspond to graduation 12°·5, whence as before—

$$x : 12°·5 :: 100 : 25 ; \quad x = \tfrac{1250°}{25} = 50° ; \quad i.e. \text{ B is the median.}$$

The second method of comparing two statistical groups, to which I alluded in the last paragraph but one, consists in stating the centesimal grade in the one group that corresponds with the median or any other fractional grade in the other. This, it will be remarked, is a very simple method of comparison, absolutely independent of any theory, and applicable to any statistical groups whatever, whether of physical or of mental qualities. Wherever we can sort in order, there we can apply this method. Thus, in the above examples, suppose A and B had been selected because they were equal when compared together, then we can concisely express the relative merits of the two classes to which they respectively belong, by saying that 16°·6 in the one is equal to 50° (the median) in the other.

I frequently make statistical records of form and feature, in the streets or in company, without exciting attention, by means of a fine pricker and a piece of paper. The pricker is a converted silver pencil-case, with the usual sliding piece; it is a very small one, and is attached to my watch chain. The pencil part has been taken out and replaced by a fine short needle, the open mouth of the case is covered with a hemispherical cap having a hole in the centre, and the adjustments are such that when the slide is pushed forward as far as it can go, the needle projects no more than one-tenth of an inch. If I then press it upon a piece of paper, held against the ball of my thumb, the paper is indelibly perforated with a fine hole, and the thumb is not wounded. The perforations will not be found to run into one another unless they are very numerous, and if they happen to do so now and then, it is of little consequence in a statistical inquiry. The holes are easily counted at leisure, by holding the paper against the light, and any scrap of paper will serve the purpose. It will be found that the majority of inquiries take the form of "more," "equal to," or "less," so I arrange the paper in a way to present three distinct compartments to the pricker, and to permit of its being held in the correct position and used by the sense of touch alone. I do so by tearing the paper into the form of a cross— that is, maimed in one of its arms—and hold it by the maimed part between the thumb and finger, the head of the cross pointing upward. The head of the cross receives the pricks referring to "more"; the solitary arm that is not maimed, those meaning "the same"; the long foot of the cross those meaning "less." It is well to write the subject of the measurement on the paper before beginning to use it, then more than one set of records can be kept in the pocket at the same time, and be severally added to as occasion serves, without fear of mistaking one for the other.

CHARACTER.

The fundamental and intrinsic differences of character that exist in individuals are well illustrated by those that distinguish the two sexes, and which begin to assert themselves even in the nursery, where all the children are treated alike. One notable peculiarity in the character of the woman is that she is capricious and coy, and has less straightforwardness than the man. It is the same in the female of every sex about the time of pairing, and there can be little doubt as to the origin of the peculiarity. If any race of animals existed in whom the sexual passions of the female were as quickly and as directly stirred as those of the male, each would mate with the first who approached her, and one essential condition of sexual selection would be absent. There would be no more call for competition among the males for the favour of each female; no more fighting for love, in which the strongest male conquers; no more rival display of personal charms, in which the best-looking or best-mannered prevails. The drama of courtship, with its prolonged strivings and doubtful success, would be cut quite short, and the race would degenerate through the absence of that sexual selection for which the protracted preliminaries of love-making give opportunity. The willy-nilly disposition of the female in matters of love is as apparent in the butterfly as in the man, and must have been continuously favoured from the earliest stages of animal evolution down to the present time. It is the factor in the great theory of sexual selection that corresponds to the insistence and directness of the male. Coyness and caprice have in consequence become a heritage of the sex, together with a cohort of allied weaknesses and petty deceits, that men have come to think venial and even amiable in women, but which they would not tolerate among themselves.

Various forms of natural character and temperament would no doubt be found to occur in constant proportions among any large group of persons of the same race, but what those proportions may be has never yet been investigated. It is extremely difficult to estimate it by observations of adults, owing to their habit of restraining natural ill tendencies, and to their long-practised concealment of those they do not restrain but desire to hide. The necessary observations ought, however, to be easily made on young children in schools, whose manifestations of character are conspicuous, who are simultaneously for months and years under the eye of the same master or mistress, and who are daily classed according to their various merits. I have occasionally asked the opinion of persons well qualified to form them, and who have had experience of teaching, as to the most obvious divisions of character to be found among school children. The replies have differed, but those on which most stress was laid were connected with energy, sociability, desire to attract notice, truthfulness, thoroughness, and refinement.

The varieties of the emotional constitution and of likings and antipathies are very numerous and wide. I may give two instances which I have not seen elsewhere alluded to, merely as examples of variation. One of them was often brought to my notice at the time when the public were admitted to see the snakes fed at the Zoological Gardens. Rabbits, birds, and other small animals were dropped in the different cages, which the snakes, after more or less serpentine action, finally struck with their poison fangs or crushed in their folds. I found it a horrible but a fascinating scene. We lead for the most part such an easy and carpeted existence, screened from the stern realities of life and death, that many of us are impelled to draw aside the curtain now and then, and gaze for a while behind it. This exhibition of the snakes at their feeding-time, which gave to me, as it doubtless did to several others, a sense of curdling of the blood, had no such effect on many of the visitors. I have often seen people—nurses, for instance, and children of all ages—looking unconcernedly and amusedly at the scene. Their indifference was perhaps the most painful element of the whole transaction. Their sympathies were absolutely unawakened. I quote this instance, partly because it leads to another very curious fact that I have noticed as regards the way with which different persons and races regard snakes. I myself have a horror of them, and can only by great self-control, and under a sense of real agitation, force myself to touch one. A considerable proportion of the English race would feel much as I do; but the remainder do not. I have questioned numbers of persons of both sexes, and have been astonished at the frequency with which I have been assured that they had no shrinking whatever from the sight of the wriggling mysterious reptile. Some persons, as is well known, make pets of them; moreover, I am told that there is no passage in Greek or Latin authors expressive of that form of horror which I myself feel, and which may be compared to what is said to be felt by hydrophobic sufferers at the undulating movements of water. There are numerous allusions in the classics to the venom fang or the crushing power of snakes, but not to an aversion inspired by its form and movement. It was the Greek symbol of Hippocrates and of healing. There is nothing of the kind in Hebrew literature, where the snake is figured as an attractive tempter. In Hindu fables the cobra is the ingenious and intelligent animal, corresponding to the fox in ours. Serpent worship was very widely spread. I therefore doubt whether the antipathy to the snake is very common among mankind, notwithstanding the instinctive terror that their sight inspires in monkeys.

The other instance I may adduce is that of the horror of blood which is curiously different in animals of the same species and in the same animals at different times. I have had a good deal of experience of the behaviour of oxen at the sight of blood, and found it to be by no means uniform. In my South African travels I relied chiefly on half-wild slaughter oxen to feed my large party, and occasionally had to shoot one on every second day. Usually the rest of the drove paid no particular heed to the place of blood, but at other rare times they seemed maddened and performed a curious sort of war-dance at the spot, making buck-leaps, brandishing their horns, and goring at the ground. It was a grotesque proceeding, utterly unlike the usual behaviour of cattle. I only witnessed it once elsewhere, and that was in the Pyrenees, where I came on a herd that was being driven homewards. Each cow in turn, as it passed a particular spot, performed the well-remembered antics. I asked, and learned that a cow had been killed there by a bear a few days previously. The natural horror at blood, and it may be the consequent dislike of red, is common among mankind; but I have seen a well-dressed child of about four years old poking its finger with a pleased innocent look into the bleeding carcase of a sheep hung up in a butcher's shop, while its nurse was inside.

The subject of character deserves more statistical investigation than it has yet received, and none have a better chance of doing it well than schoolmasters; their opportunities are indeed most enviable. It would be necessary to approach the subject wholly without prejudice, as a pure matter of observation, just as if the children were the fauna and flora of hitherto undescribed species in an entirely new land.

CRIMINALS AND THE INSANE.

Criminality, though not very various in its development, is extremely complex in its origin; nevertheless certain general conclusions are arrived at by the best writers on the subject, among whom Prosper Despine is one of the most instructive. The ideal criminal has marked peculiarities of character: his conscience is almost deficient, his instincts are vicious, his power of self-control is very weak, and he usually detests continuous labour. The absence of self-control is due to ungovernable temper, to passion, or to mere imbecility, and the conditions that determine the particular description of crime are the character of the instincts and of the temptation.

The deficiency of conscience in criminals, as shown by the absence of genuine remorse for their guilt, astonishes all who first become familiar with the details of prison life. Scenes of heartrending despair are hardly ever witnessed among prisoners; their sleep is broken by no uneasy dreams —on the contrary, it is easy and sound; they have also excellent appetites. But hypocrisy is a very common vice; and all my information agrees as to the utter untruthfulness of criminals, however plausible their statements may be.

We must guard ourselves against looking upon vicious instincts as perversions, inasmuch as they may be strictly in accordance with the healthy nature of the man, and, being transmissible by inheritance, may become the normal characteristics of a healthy race, just as the sheep-dog, the retriever, the pointer, and the bull-dog, have their several instincts. There can be no greater popular error than the supposition that natural instinct is a perfectly trustworthy guide, for there are striking contradictions to such an opinion in individuals of every description of animal. The most that we are entitled to say in any case is, that the prevalent instincts of each race are trustworthy, not those of every individual. But even this is saying too much, because when the conditions under which the race is living have recently been changed, some instincts which were adapted to the old state of things are sure to be fallacious guides to conduct in the new one. A man who is counted as an atrocious criminal in England, and is punished as such by English law in social self-defence, may nevertheless have acted in strict accordance with instincts that are laudable in less civilised societies. The ideal criminal is, unhappily for him, deficient in qualities that are capable of restraining his unkindly or inconvenient instincts ; he has neither sympathy for others nor the sense of duty, both of which lie at the base of conscience ; nor has he sufficient self-control to accommodate himself to the society in which he has to live, and so to promote his own selfish interests in the long-run. He cannot be preserved from criminal misadventure, either by altruistic sentiments or by intelligently egoistic ones.

The perpetuation of the criminal class by heredity is a question difficult to grapple with on many accounts. Their vagrant habits, their illegitimate unions, and extreme untruthfulness, are among the difficulties of the investigation. It is, however, easy to show that the criminal nature tends to be inherited ; while, on the other hand, it is impossible that women who spend a large portion of the best years of their life in prison can contribute many children to the population. The true state of the case appears to be that the criminal population receives steady accessions from those who, without having strongly-marked criminal natures, do nevertheless belong to a type of humanity that is exceedingly ill suited to play a respectable part in our modern civilisation, though it is well suited to flourish under half-savage conditions, being naturally both healthy and prolific. These persons are apt to go to the bad ; their daughters consort with criminals and become the parents of criminals. An extraordinary example of this is afforded by the history of the infamous Jukes family in America, whose pedigree has been made out, with extraordinary care, during no less than seven generations, and is the subject of an elaborate memoir printed in the Thirty-first Annual Report of the Prison Association of New York, 1876.   It includes no less than 540 individuals of Jukes blood, of whom a frightful number degraded into criminality, pauperism, or disease.

It is difficult to summarise the results in a few plain figures, but I will state those respecting the fifth generation, through the eldest of the five prolific daughters of the man who is the common ancestor of the race.   The total number of these was 123, of whom thirty-eight came through an illegitimate granddaughter, and eighty-five through legitimate grandchildren.   Out of the thirty-eight, sixteen have been in jail, six of them for heinous offences, one of these having been committed no less than nine times; eleven others led openly disreputable lives or were paupers; four were notoriously intemperate; the history of three had not been traced, and only four are known to have done well. The great majority of the women consorted with criminals. As to the eighty-five legitimate descendants, they were less flagrantly bad, for only five of them had been in jail, and only thirteen others had been paupers.   Now the ancestor of all this mischief, who was born about the year 1730, is described as having been a jolly companionable man, a hunter, and a fisher, averse to steady labour, but working hard and idling by turns, and who had numerous illegitimate children, whose issue has not been traced.   He was, in fact, a somewhat good specimen of a half-savage, without any seriously criminal instincts.   The girls were apparently attractive, marrying early and sometimes not badly ; but the gipsy-like character of the race was unsuited to success in a civilised country.   So the descendants went to the bad, and such hereditary moral weaknesses as they may have had, rose to the surface and worked their mischief without check.   Cohabiting with criminals, and being extremely prolific, the result was the production of a stock exceeding 500 in number, of a prevalent criminal type.   Through disease and intemperance the breed is now rapidly diminishing; the infant mortality has of late been horrible, but fortunately the women of the present generation bear usually but few children, and many of them are altogether childless.

The criminal classes contain a considerable portion of epileptics and other persons of instable, emotional temperament, subject to nervous explosions that burst out at intervals and relieve the system. The mad outbreaks of women in convict prisons is a most curious phenomenon. Some of them are apt from time to time to have a gradually increasing desire that at last becomes irresistible, to "break out," as it is technically called ; that is, to smash and tear everything they can within reach, and to shriek, curse, and howl. At length the fit expends itself ; the devil, as it were, leaves them, and they begin to behave again in their ordinary way. The highest form of emotional instability exists in confirmed epilepsy, where its manifestations have often been studied; it is found in a high but somewhat less extraordinary degree in the hysterical and allied affections. In the confirmed epileptic constitution the signs of general instability of nervous action are muscular convulsions, irregularities of bodily temperature, mobile intellectual activity, and extraordinary oscillations between opposed emotional states. I am assured by excellent authority that instable manifestations of extreme piety and of extreme vice are almost invariably shown by epileptics, and should be regarded as a prominent feature of their peculiar constitution. These unfortunate beings see no incongruity between the pious phrases that they pour out at one moment and their vile and obscene language in the next ; neither do they show repentance for past misconduct when they are convicted of crimes, however abominable these may be. They are creatures of the moment, possessing no inhibitory check upon their desires and emotions, which drive them headlong hither and thither.

Madness is often associated with epilepsy ; in all cases it is a frightful and hereditary disfigurement of humanity, which appears, from the upshot of various conflicting accounts, to be on the increase. The neurotic constitution.

from which it springs is however not without its merits, as has been well pointed out, since a large proportion of the enthusiastic men and women to whose labour the world is largely indebted, have had that constitution, judging from the fact that insanity existed in their families.

The phases of extreme piety and extreme vice which so rapidly succeed one another in the same individual among the epileptics, are more widely separated among those who are simply insane. It has been noticed that among the morbid organic conditions which accompany the show of excessive piety and religious rapture in the insane, none are so frequent as disorders of the sexual organisation. Conversely, the frenzies of religious revivals have not unfrequently ended in gross profligacy. The encouragement of celibacy by the fervent leaders of most creeds, utilises in an unconscious way the morbid connection between an over-restraint of the sexual desires and impulses towards extreme devotion.

Another remarkable phase among the insane consists in strange views about their individuality. They think that their body is made of glass, or that their brains have literally disappeared, or that there are different persons inside them, or that they are somebody else, and so forth. It is said that this phase is most commonly associated with morbid disturbance of the alimentary organs. So in many religions fasting has been used as an agent for detaching the thoughts from the body and for inducing ecstasy.

There is yet a third peculiarity of the insane which is almost universal, that of gloomy segregation. Passengers nearing London by the Great Western Railway must have frequently remarked the unusual appearance of the crowd of lunatics when taking their exercise in the large green enclosure in front of Hanwell Asylum. They almost without exception walk apart in moody isolation, each in his own way, buried in his own thoughts. It is a scene like that fabled in Vathek's hall of Eblis. I am assured that whenever two are seen in company, it is either because their attacks of madness are of an intermittent and epileptic character and they are temporarily sane, or otherwise that they are near recovery. Conversely, the curative influence of social habits is fully recognised, and they are promoted

by festivities in the asylums.   On the other hand, the great
teachers of all creeds have made seclusion a prominent
religious exercise.   In short, by enforcing celibacy, fasting,
and solitude, they have done their best towards making men
mad, and they have always largely succeeded in inducing
morbid mental conditions among their followers.

Floods of light are thrown upon various incidents of
devotee life, and also upon the disgusting and not otherwise intelligible character of the sanctimonious scoundrel,
by the everyday experiences of the madhouse.   No professor of metaphysics, psychology, or religion can claim
to know the elements of what he teaches, unless he is
acquainted with the ordinary phenomena of idiocy, madness, and epilepsy.   He must study the manifestations of
disease and congenital folly, as well as those of sanity and
high intellect.

Gregarious and Slavish Instincts.

I propose in this chapter to discuss a curious and apparently anomalous group of base moral instincts and intellectual deficiencies, that are innate rather than acquired,
by tracing their analogies in the world of brutes and
examining the conditions through which they have been
evolved.   They are the slavish aptitudes from which the
leaders of men are exempt, but which are characteristic
elements in the disposition of ordinary persons.   The vast
majority of persons of our race have a natural tendency to
shrink from the responsibility of standing and acting alone ;
they exalt the *vox populi*, even when they know it to be the
utterance of a mob of nobodies, into the *vox Dei*, and they
are willing slaves to tradition, authority, and custom.   The
intellectual deficiencies corresponding to these moral flaws
are shown by the rareness of free and original thought as
compared with the frequency and readiness with which men
accept the opinions of those in authority as binding on their
judgment.   I shall endeavour to prove that the slavish
aptitudes in man are a direct consequence of his gregarious
nature, which itself is a result of the conditions both of his
primeval barbarism and of the forms of his subsequent
civilisation.   My argument will be, that gregarious brute

animals possess a want of self-reliance in a marked degree; that the conditions of the lives of these animals have made a want of self-reliance a necessity to them, and that by the law of natural selection the gregarious instincts and their accompanying slavish aptitudes have gradually become evolved. Then I shall argue that our remote ancestors have lived under parallel conditions, and that other causes peculiar to human society have acted up to the present day in the same direction, and that we have inherited the gregarious instincts and slavish aptitudes which have been needed under past circumstances, although in our advancing civilisation they are becoming of more harm than good to our race.

It was my fortune, in earlier life, to gain an intimate knowledge of certain classes of gregarious animals. The urgent need of the camel for the close companionship of his fellows was a never-exhausted topic of curious admiration to me during tedious days of travel across many North African deserts. I also happened to hear and read a great deal about the still more marked gregarious instincts of the llama; but the social animal into whose psychology I am conscious of having penetrated most thoroughly is the ox of the wild parts of western South Africa. It is necessary to insist upon the epithet "wild," because an ox of tamed parentage has different natural instincts; for instance, an English ox is far less gregarious than those I am about to describe, and affords a proportionately less valuable illustration to my argument. The oxen of which I speak belonged to the Damaras, and none of the ancestry of these cattle had ever been broken to harness. They were watched from a distance during the day, as they roamed about the open country, and at night they were driven with cries to enclosures, into which they rushed much like a body of terrified wild animals driven by huntsmen into a trap. Their scared temper was such as to make it impossible to lay hold of them by other means than by driving the whole herd into a clump, and lassoing the leg of the animal it was desired to seize, and throwing him to the ground with dexterous force. With oxen and cows of this description, whose nature is no doubt shared by the bulls, I spent more than a year in the closest companionship.

I had nearly a hundred of the beasts broken in for the waggon, for packs, and for the saddle. I travelled an entire journey of exploration on the back of one of them, with others by my side, either labouring at their tasks or walking at leisure; and with others again who were wholly unbroken, and who served the purpose of an itinerant larder. At night, when there had been no time to erect an enclosure to hold them, I lay down in their midst, and it was interesting to observe how readily they then availed themselves of the neighbourhood of the camp fire and of man, conscious of the protection they afforded from prowling carnivora, whose cries and roars, now distant, now near, continually broke upon the stillness. These opportunities of studying the disposition of such peculiar cattle were not wasted upon me. I had only too much leisure to think about them, and the habits of the animals strongly attracted my curiosity. The better I understood them, the more complex and worthy of study did their minds appear to be. But I am now concerned only with their blind gregarious instincts, which are conspicuously distinct from the ordinary social desires. In the latter they are deficient; thus they are not amiable to one another, but show on the whole more expressions of spite and disgust than of forbearance or fondness. They do not suffer from an ennui, which society can remove, because their coarse feeding and their ruminant habits make them somewhat stolid. Neither can they love society, as monkeys do, for the opportunities it affords of a fuller and more varied life, because they remain self-absorbed in the middle of their herd, while the monkeys revel together in frolics, scrambles, fights, loves, and chatterings. Yet although the ox has so little affection for, or individual interest in, his fellows, he cannot endure even a momentary severance from his herd. If he be separated from it by stratagem or force, he exhibits every sign of mental agony; he strives with all his might to get back again, and when he succeeds, he plunges into its middle to bathe his whole body with the comfort of closest companionship. This passionate terror at segregation is a convenience to the herdsman, who may rest assured in the darkness or in the mist that the whole herd is safe whenever he can get a glimpse of a single ox. It is also the cause of great inconvenience to the traveller in ox-waggons, who constantly feels himself in a position towards his oxen like that of a host to a company of bashful gentlemen at the time when he is trying to get them to move from the drawing-room to the dinner-table, and no one will go first, but every one backs and gives place to his neighbour. The traveller finds great difficulty in procuring animals capable of acting the part of fore-oxen to his team, the ordinary members of the wild herd being wholly unfitted by nature to move in so prominent and isolated a position, even though, as is the custom, a boy is always in front to persuade or pull them onwards. Therefore, a good fore-ox is an animal of an exceptionally independent disposition. Men who break in wild cattle for harness watch assiduously for those who show a self-reliant nature, by grazing apart or ahead of the rest, and these they break in for fore-oxen. The other cattle may be indifferently devoted to ordinary harness purposes, or to slaughter ; but the born leaders are far too rare to be used for any less distinguished service than that which they alone are capable of fulfilling. But a still more exceptional degree of merit may sometimes be met with among the many thousands of Damara cattle. It is possible to find an ox who may be ridden, not indeed as freely as a horse, for I have never heard of a feat like that, but at all events wholly apart from the companionship of others ; and an accomplished rider will even succeed in urging him out at a trot from the very middle of his fellows. With respect to the negative side of the scale, though I do not recollect definite instances, I can recall general impressions of oxen showing a deficiency from the average ox standard of self-reliance, about equal to the excess of that quality found in ordinary fore-oxen. Thus I recollect there were some cattle of a peculiarly centripetal instinct, who ran more madly than the rest into the middle of the herd when they were frightened ; and I have no reason to doubt from general recollections that the law of deviation from an average would be as applicable to independence of character among cattle as one might expect it theoretically to be. The conclusion to which we are driven is, that few of the Damara cattle have enough originality and independence of disposition to pass unaided through their daily risks in a tolerably comfortable manner. They are essentially slavish, and seek no better lot than to be led by any one of their number who has enough self-reliance to accept that position. No ox ever dares to act contrary to the rest of the herd, but he accepts their common determination as an authority binding on his conscience.

An incapacity of relying on oneself and a faith in others are precisely the conditions that compel brutes to congregate and live in herds; and, again, it is essential to their safety in a country infested by large carnivora, that they should keep closely together in herds. No ox grazing alone could live for many days unless he were protected, far more assiduously and closely than is possible to barbarians. The Damara owners confide perhaps 200 cattle to a couple of half-starved youths, who pass their time in dozing or in grubbing up roots to eat. The owners know that it is hopeless to protect the herd from lions, so they leave it to take its chance; and as regards human marauders they equally know that the largest number of cattle watchers they could spare could make no adequate resistance to an attack; they therefore do not send more than two, who are enough to run home and give the alarm to the whole male population of the tribe to run in arms on the tracks of their plundered property. Consequently, as I began by saying, the cattle have to take care of themselves against the wild beasts, and they would infallibly be destroyed by them if they had not safeguards of their own, which are not easily to be appreciated at first sight at their full value. We shall understand them better by considering the precise nature of the danger that an ox runs. When he is alone it is not simply that he is too defenceless, but that he is easily surprised. A crouching lion fears cattle who turn boldly upon him, and he does so with reason. The horns of an ox or antelope are able to make an ugly wound in the paw or chest of a springing beast when he receives its thrust in the same way that an over-eager pugilist meets his adversary's "counter" hit. Hence it is that a cow who has calved by the wayside, and has been temporarily abandoned by the caravan, is never seized by lions. The incident frequently occurs, and as frequently are the cow and calf eventually brought safe to the camp; and yet there is usually evidence in footprints of her having sustained a regular siege from the wild beasts; but she is so restless and eager for the safety of her young that no beast of prey can approach her unawares. This state of exaltation is of course exceptional; cattle are obliged in their ordinary course of life to spend a considerable part of the day with their heads buried in the grass, where they can neither see nor smell what is about them. A still larger part of their time must be spent in placid rumination, during which they cannot possibly be on the alert. But a herd of such animals, when considered as a whole, is always on the alert; at almost every moment some eyes, ears, and noses will command all approaches, and the start or cry of alarm of a single beast is a signal to all his companions. To live gregariously is to become a fibre in a vast sentient web overspreading many acres; it is to become the possessor of faculties always awake, of eyes that see in all directions, of ears and nostrils that explore a broad belt of air; it is also to become the occupier of every bit of vantage ground whence the approach of a wild beast might be overlooked. The protective senses of each individual who chooses to live in companionship are multiplied by a large factor, and he thereby receives a maximum of security at a minimum cost of restlessness. When we isolate an animal who has been accustomed to a gregarious life, we take away his sense of protection, for he feels himself exposed to danger from every part of the circle around him, except the one point on which his attention is momentarily fixed; and he knows that disaster may easily creep up to him from behind. Consequently his glance is restless and anxious, and is turned in succession to different quarters; his movements are hurried and agitated, and he becomes a prey to the extremest terror. There can be no room for doubt that it is suitable to the well-being of cattle in a country infested with beasts of prey to live in close companionship, and being suitable, it follows from the law of natural selection that the development of gregarious and therefore of slavish instincts must be favoured in such cattle. It also follows from the same law that the degree in which those instincts are developed is on the whole the most conducive to their safety. If they were more gregarious they would crowd so closely as to interfere with each other when grazing the scattered pasture of Damara land; if less gregarious, they would be too widely scattered to keep a sufficient watch against the wild beasts.

I now proceed to consider more particularly why the range of deviation from the average is such that we find about one ox out of fifty to possess sufficient independence of character to serve as a pretty good fore-ox. Why is it not one in five or one in five hundred? The reason undoubtedly is that natural selection tends to give but one leader to each suitably-sized herd, and to repress superabundant leaders. There is a certain size of herd most suitable to the geographical and other conditions of the country; it must not be too large, or the scattered puddles which form their only watering-places for a great part of the year would not suffice; and there are similar drawbacks in respect to pasture. It must not be too small, or it would be comparatively insecure; thus a troop of five animals is far more easy to be approached by a stalking huntsman than one of twenty, and the latter than one of a hundred. We have seen that it is the oxen who graze apart, as well as those who lead the herd, who are recognised by the trainers of cattle as gifted with enough independence of character to become fore-oxen. They are even preferred to the actual leaders of the herd; they dare to move more alone, and therefore their independence is undoubted. The leaders are safe enough from lions, because their flanks and rear are guarded by their followers; but each of those who graze apart, and who represent the superabundant supply of self-reliant animals, have one flank and the rear exposed, and it is precisely these whom the lions take. Looking at the matter in a broad way, we may justly assert that wild beasts trim and prune every herd into compactness, and tend to reduce it into a closely-united body with a single well-protected leader. That the development of independence of character in cattle is thus suppressed below its otherwise natural standard by the influence of wild beasts, is shown by the greater display of self-reliance among cattle whose ancestry for some generations have not been exposed to such danger.

What has been said about cattle, in relation to wild

beasts, applies with more or less obvious modifications to barbarians in relation to their neighbours, but I insist on a close resemblance in the particular circumstance, that many savages are so unamiable and morose as to have hardly any object in associating together, besides that of mutual support.   If we look at the inhabitants of the very same country as the oxen I have described, we shall find them congregated into multitudes of tribes, all more or less at war with one another.   We shall find that few of these tribes are very small, and few very large, and that it is precisely those that are exceptionally large or small whose condition is the least stable.   A very small tribe is sure to be overthrown, slaughtered, or driven into slavery by its more powerful neighbour.   A very large tribe falls to pieces through its own unwieldiness, because, by the nature of things, it must be either deficient in centralisation or straitened in food, or both.   A barbarian population is obliged to live dispersedly, since a square mile of land will support only a few hunters or shepherds ; on the other hand, a barbarian government cannot be long maintained unless the chief is brought into frequent contact with his dependants, and this is geographically impossible when his tribe is so scattered as to cover a great extent of territory.   The law of selection must discourage every race of barbarians which supplies self-reliant individuals in such large numbers as to cause tribes of moderate size to lose their blind desire of aggregation. It must equally discourage a breed that is incompetent to supply such men in sufficiently abundant ratio to the rest of the population to ensure the existence of tribes of not too large a size.   It must not be supposed that gregarious instincts are equally important to all forms of savage life ; but I hold, from what we know of the clannish fighting habits of our forefathers, that they were every whit as applicable to the earlier ancestors of our European stock as they are still to a large part of the black population of Africa.

There is, moreover, an extraordinary power of tyranny invested in the chiefs of tribes and nations of men, that so vastly outweighs the analogous power possessed by the leaders of animal herds as to rank as a special attribute of human society, eminently conducive to slavishness.   If any brute in a herd makes itself obnoxious to the leader, the leader attacks him, and there is a free fight between the two, the other animals looking on the while. But if a man makes himself obnoxious to his chief, he is attacked, not by the chief single-handed, but by the overpowering force of his executive. The rebellious individual has to brave a disciplined host; there are spies who will report his doings, a local authority who will send a detachment of soldiers to drag him to trial; there are prisons ready built to hold him, civil authorities wielding legal powers of stripping him of all his possessions, and official executioners prepared to torture or kill him. The tyrannies under which men have lived, whether under rude barbarian chiefs, under the great despotisms of half-civilised Oriental countries, or under some of the more polished but little less severe governments of modern days, must have had a frightful influence in eliminating independence of character from the human race. Think of Austria, of Naples, and even of France under Napoleon III. It was stated [1] in 1870 that, according to papers found at the Tuileries, 26,642 persons had been arrested in France for political offences since 2nd December, 1851, and that 14,118 had been transported, exiled, or detained in prison.

I have already spoken in *Hereditary Genius* of the large effects of religious persecution in comparatively recent years, on the natural character of races, and shall not say more about it here ; but it must not be omitted from the list of steady influences continuing through ancient historical times down, in some degree, to the present day, in destroying the self-reliant, and therefore the nobler races of men.

I hold that the blind instincts evolved under these long-continued conditions have been ingrained into our breed, and that they are a bar to our enjoying the freedom which the forms of modern civilisation are otherwise capable of giving us. A really intelligent nation might be held together by far stronger forces than are derived from the purely gregarious instincts. A nation need not be a mob of slaves, clinging to one another through fear, and for the most part incapable of self-government, and begging to be led ; but it might consist of vigorous self-reliant men, knit to one

[1] *Daily News*, 17th October, 1870.

another by innumerable ties, into a strong, tense, and elastic organisation.

The character of the corporate action of a nation in which each man judges for himself, might be expected to possess statistical constancy. It would be the expression of the dominant character of a large number of separate members of the same race, and ought therefore to be remarkably uniform. Fickleness of national character is principally due to the several members of the nation exercising no independent judgment, but allowing themselves to be led hither and thither by the successive journalists, orators, and sentimentalists who happen for the time to have the chance of directing them.

Our present natural dispositions make it impossible for us to attain the ideal standard of a nation of men all judging soberly for themselves, and therefore the slavishness of the mass of our countrymen, in morals and intellect, must be an admitted fact in all schemes of regenerative policy.

The hereditary taint due to the primeval barbarism of our race, and maintained by later influences, will have to be bred out of it before our descendants can rise to the position of free members of an intelligent society: and I may add that the most likely nest at the present time for self-reliant natures is to be found in States founded and maintained by emigrants.

Servility has its romantic side, in the utter devotion of a slave to the lightest wishes and the smallest comforts of his master, and in that of a loyal subject to those of his sovereign; but such devotion cannot be called a reasonable self-sacrifice; it is rather an abnegation of the trust imposed on man to use his best judgment, and to act in the way he thinks the wisest. Trust in authority is a trait of the character of children, of weakly women, and of the sick and infirm, but it is out of place among members of a thriving resolute community during the fifty or more years of their middle life. Those who have been born in a free country feel the atmosphere of a paternal government very oppressive. The hearty and earnest political and individual life which is found when every man has a continual sense of public responsibility, and knows that success depends on his own right judgment and exertion, is replaced under a despotism by an indolent reliance upon what its master may direct, and by a demoralising conviction that personal advancement is best secured by solicitations and favour.

INTELLECTUAL DIFFERENCES.

It is needless for me to speak here about the differences in intellectual power between different men and different races, or about the convertibility of genius as shown by different members of the same gifted family achieving eminence in varied ways, as I have already written at length on these subjects in *Hereditary Genius* and in *Antecedents of English Men of Science.* It is, however, well to remark that during the fourteen years that have elapsed since the former book was published, numerous fresh instances have arisen of distinction being attained by members of the gifted families whom I quoted as instances of heredity, thus strengthening my arguments.

MENTAL IMAGERY.

Anecdotes find their way into print, from time to time, of persons whose visual memory is so clear and sharp as to present mental pictures that may be scrutinised with nearly as much ease and prolonged attention as if they were real objects. I became interested in the subject and made a rather extensive inquiry into the mode of visual presentation in different persons, so far as could be gathered from their respective statements. It seemed to me that the results might illustrate the essential differences between the mental operations of different men, that they might give some clue to the origin of visions, and that the course of the inquiry might reveal some previously unnoticed facts. It has done all this more or less, and I will explain the results in the present and in the three following chapters.

It is not necessary to trouble the reader with my earlier tentative steps to find out what I desired to learn. After the inquiry had been fairly started it took the form of submitting a certain number of printed questions to a large number of

persons (see Appendix E).   There is hardly any more difficult task than that of framing questions which are not likely
to be misunderstood, which admit of easy reply, and which
cover the ground of inquiry.   I did my best in these respects, without forgetting the most important part of all—
namely, to tempt my correspondents to write freely in fuller
explanation of their replies, and on cognate topics as well.
These separate letters have proved more instructive and
interesting by far than the replies to the set questions.

The first group of the rather long series of queries related
to the illumination, definition, and colouring of the mental
image, and were framed thus :—

" Before addressing yourself to any of the Questions on the
opposite page, think of some definite object—suppose it is your
breakfast-table as you sat down to it this morning—and consider carefully the picture that rises before your mind's eye.

1. *Illumination.*—Is the image dim or fairly clear? Is its
brightness comparable to that of the actual scene ?

2. *Definition.*—Are all the objects pretty well defined at the
same time, or is the place of sharpest definition at any one
moment more contracted than it is in a real scene?

3. *Colouring.*—Are the colours of the china, of the toast,
bread-crust, mustard, meat, parsley, or whatever may have been
on the table, quite distinct and natural ? "

The earliest results of my inquiry amazed me.   I had
begun by questioning friends in the scientific world, as they
were the most likely class of men to give accurate answers
concerning this faculty of visualising, to which novelists and
poets continually allude, which has left an abiding mark
on the vocabularies of every language, and which supplies
the material out of which dreams and the well-known
hallucinations of sick people are built.

To my astonishment, I found that the great majority of
the men of science to whom I first applied protested that
mental imagery was unknown to them, and they looked on
me as fanciful and fantastic in supposing that the words
" mental imagery " really expressed what I believed everybody supposed them to mean.   They had no more notion
of its true nature than a colour-blind man, who has not
discerned his defect, has of the nature of colour.   They
had a mental deficiency of which they were unaware, and

naturally enough supposed that those who affirmed they possessed it, were romancing. To illustrate their mental attitude it will be sufficient to quote a few lines from the letter of one of my correspondents, who writes :—

"These questions presuppose assent to some sort of a proposition regarding the 'mind's eye,' and the 'images' which it sees. . . . This points to some initial fallacy. . . . It is only by a figure of speech that I can describe my recollection of a scene as a 'mental image' which I can 'see' with my 'mind's eye.' . . . I do not see it . . . any more than a man sees the thousand lines of Sophocles which under due pressure he is ready to repeat. The memory possesses it, etc."

Much the same result followed inquiries made for me by a friend among members of the French Institute.

On the other hand, when I spoke to persons whom I met in general society, I found an entirely different disposition to prevail. Many men and a yet larger number of women, and many boys and girls, declared that they habitually saw mental imagery, and that it was perfectly distinct to them and full of colour. The more I pressed and cross-questioned them, professing myself to be incredulous, the more obvious was the truth of their first assertions. They described their imagery in minute detail, and they spoke in a tone of surprise at my apparent hesitation in accepting what they said. I felt that I myself should have spoken exactly as they did if I had been describing a scene that lay before my eyes, in broad daylight, to a blind man who persisted in doubting the reality of vision. Reassured by this happier experience, I recommenced to inquire among scientific men, and soon found scattered instances of what I sought, though in by no means the same abundance as elsewhere. I then circulated my questions more generally among my friends and through their hands, and obtained the replies that are the main subject of this and of the three next chapters. They were from persons of both sexes, and of various ages, and in the end from occasional correspondents in nearly every civilised country.

I have also received batches of answers from various educational establishments both in England and America, which were made after the masters had fully explained the meaning of the questions, and interested the boys in them. These have the merit of returns derived from a general census, which my other data lack, because I cannot for a moment suppose that the writers of the latter are a haphazard proportion of those to whom they were sent. Indeed I know of some who, disavowing all possession of the power, and of many others who, possessing it in too faint a degree to enable them to express what their experiences really were, in a manner satisfactory to themselves, sent no returns at all. Considerable statistical similarity was, however, observed between the sets of returns furnished by the schoolboys and those sent by my separate correspondents, and I may add that they accord in this respect with the oral information I have elsewhere obtained. The conformity of replies from so many different sources which was clear from the first, the fact of their apparent trustworthiness being on the whole much increased by cross-examination (though I could give one or two amusing instances of break-down), and the evident effort made to give accurate answers, have convinced me that it is a much easier matter than I had anticipated to obtain trustworthy replies to psychological questions. Many persons, especially women and intelligent children, take pleasure in introspection, and strive their very best to explain their mental processes. I think that a delight in self-dissection must be a strong ingredient in the pleasure that many are said to take in confessing themselves to priests.

Here, then, are two rather notable results : the one is the proved facility of obtaining statistical insight into the processes of other persons' minds, whatever à priori objection may have been made as to its possibility ; and the other is that scientific men, as a class, have feeble powers of visual representation. There is no doubt whatever on the latter point, however it may be accounted for. My own conclusion is, that an over-ready perception of sharp mental pictures is antagonistic to the acquirement of habits of highly-generalised and abstract thought, especially when the steps of reasoning are carried on by words as symbols, and that if the faculty of seeing the pictures was ever possessed by men who think hard, it is very apt to be lost by disuse. The highest minds are probably those in which it is not lost, but subordinated, and is ready for use on suitable occasions. I am, however, bound to say, that the missing faculty seems to be replaced so serviceably by other modes of conception, chiefly, I believe, connected with the incipient motor sense, not of the eyeballs only but of the muscles generally, that men who declare themselves entirely deficient in the power of seeing mental pictures can nevertheless give life-like descriptions of what they have seen, and can otherwise express themselves as if they were gifted with a vivid visual imagination. They can also become painters of the rank of Royal Academicians.

The facts I am now about to relate are obtained from the returns of 100 adult men, of whom 19 are Fellows of the Royal Society, mostly of very high repute, and at least twice, and I think I may say three times, as many more are persons of distinction in various kinds of intellectual work. As already remarked, these returns taken by themselves do not profess to be of service in a general statistical sense, but they are of much importance in showing how men of exceptional accuracy express themselves when they are speaking of mental imagery. They also testify to the variety of experiences to be met with in a moderately large circle. I will begin by giving a few cases of the highest, of the medium, and of the lowest order of the faculty of visualising. The hundred returns were first classified according to the order of the faculty, as judged to the best of my ability from the whole of what was said in them, and of what I knew from other sources of the writers; and the number prefixed to each quotation shows its place in the class-list.

VIVIDNESS OF MENTAL IMAGERY.

(From returns, furnished by 100 men, at least half of whom are distinguished in science or in other fields of intellectual work.)

Cases where the faculty is very high.

1. Brilliant, distinct, never blotchy.
2. Quite comparable to the real object. I feel as though I was dazzled, *e.g.* when recalling the sun to my mental vision.

3. In some instances quite as bright as an actual scene.

4. Brightness as in the actual scene.

5. Thinking of the breakfast-table this morning, all the objects in my mental picture are as bright as the actual scene.

6. The image once seen is perfectly clear and bright.

7. Brightness at first quite comparable to actual scene.

8. The mental image appears to correspond in all respects with reality. I think it is as clear as the actual scene.

9. The brightness is perfectly comparable to that of the real scene.

10. I think the illumination of the imaginary image is nearly equal to that of the real one.

11. All clear and bright ; all the objects seem to me well defined at the same time.

12. I can see my breakfast-table or any equally familiar thing with my mind's eye, quite as well in all particulars as I can do if the reality is before me.

Cases where the faculty is mediocre.

46. Fairly clear and not incomparable in illumination with that of the real scene, especially when I first catch it. Apt to become fainter when more particularly attended to.

47. Fairly clear, not quite comparable to that of the actual scene. Some objects are more sharply defined than others, the more familiar objects coming more distinctly in my mind.

48. Fairly clear as a general image ; details rather misty.

49. Fairly clear, but not equal to the scene. Defined, but not sharply ; not all seen with equal clearness.

50. Fairly clear. Brightness probably at least one-half to two-thirds of original. [The writer is a physiologist.] Definition varies very much, one or two objects being much more distinct than the others, but the latter come out clearly if attention be paid to them.

51. Image of my breakfast-table fairly clear, but not quite so bright as the reality. Altogether it is pretty well defined ; the part where I sit and its surroundings are pretty well so.

52. Fairly clear, but brightness not comparable to that of the actual scene. The objects are sharply defined ; some of them are salient, and others insignificant and dim, but by separate efforts I can take a visualised inventory of the whole table.

53. Details of breakfast-table *when the scene is reflected on* are fairly defined and complete, but I have had a familiarity of many years with my own breakfast-table, and the above would not be the case with a table seen casually unless there were some striking peculiarity in it.

54. I can recall any single object or group of objects, but not

the whole table at once. The things recalled are generally clearly defined. Our table is a long one ; I can in my mind pass my eyes all down the table and see the different things distinctly, but not the whole table at once.

Cases where the faculty is at the lowest.

89. Dim and indistinct, yet I can give an account of this morning's breakfast-table ; split herrings, broiled chickens, bacon, rolls, rather light-coloured marmalade, faint green plates with stiff pink flowers, the girls' dresses, etc. etc. I can also tell where all the dishes were, and where the people sat (I was on a visit). But my imagination is seldom pictorial except between sleeping and waking, when I sometimes see rather vivid forms.

90. Dim and not comparable in brightness to the real scene. Badly defined with blotches of light ; very incomplete.

91. Dim, poor definition ; could not sketch from it. I have a difficulty in seeing two images together.

92. Usually very dim. I cannot speak of its brightness, but only of its faintness. Not well defined and very incomplete.

93. Dim, imperfect.

94. I am very rarely able to recall any object whatever with any sort of distinctness. Very occasionally an object or image will recall itself, but even then it is more like a generalised image than an individual image. I seem to be almost destitute of visualising power, as under control.

95. No power of visualising. Between sleeping and waking, in illness and in health, with eyes closed, some remarkable scenes have occasionally presented themselves, but I cannot recall them when awake with eyes open, and by daylight, or under any circumstances whatever when a copy could be made of them on paper. I have drawn both men and places many days or weeks after seeing them, but it was by an effort of memory acting on study at the time, and assisted by trial and error on the paper or canvas, whether in black, yellow, or colour, afterwards.

96. It is only as a figure of speech that I can describe my recollection of a scene as a " mental image " which I can " see " with my " mind's eye." . . . The memory possesses it, and the mind can at will roam over the whole, or study minutely any part.

97. No individual objects, only a general idea of a very uncertain kind.

98. No. My memory is not of the nature of a spontaneous vision, though I remember well where a word occurs in a page, how furniture looks in a room, etc. The ideas not felt to be mental pictures, but rather the symbols of facts.

99. Extremely dim. The impressions are in all respects so dim, vague, and transient, that I doubt whether they can reasonably be called images. They are incomparably less than those of dreams.

100. My powers are zero. To my consciousness there is almost no association of memory with objective visual impressions. I recollect the breakfast-table, but do not see it.

These quotations clearly show the great variety of natural powers of visual representation, and though the returns from which they are taken have, as I said, no claim to be those of 100 Englishmen taken at haphazard, nevertheless, to the best of my judgment, they happen to differ among themselves in much the same way that such returns would have done. I cannot procure a strictly haphazard series for comparison, because in any group of persons whom I may question there are always many too indolent to reply, or incapable of expressing themselves, or who from some fancy of their own are unwilling to reply. Still, as already mentioned, I have got together several groups that approximate to what is wanted, usually from schools, and I have analysed them as well as I could, and the general result is that the above returns may be accepted as a fair representation of the visualising powers of Englishmen. Treating these according to the method described in the chapter of statistics, we have the following results, in which, as a matter of interest, I have also recorded the highest and the lowest of the series :—

Highest.—Brilliant, distinct, never blotchy.

First Suboctile.—The image once seen is perfectly clear and bright.

First Octile.—I can see my breakfast-table or any equally familiar thing with my mind's eye quite as well in all particulars as I can do if the reality is before me.

First Quartile—Fairly clear ; illumination of actual scene is fairly represented. Well defined. Parts do not obtrude themselves, but attention has to be directed to different points in succession to call up the whole.

Middlemost.—Fairly clear. Brightness probably at least from one-half to two-thirds of the original. Definition varies very much, one or two objects being much more distinct than the others, but the latter come out clearly if attention be paid to them.

Last Quartile.—Dim, certainly not comparable to the actual scene. I have to think separately of the several things on the table to bring them clearly before the mind's eye, and when I think of some things the others fade away in confusion.

Last Octile.—Dim and not comparable in brightness to the real scene. Badly defined, with blotches of light ; very incomplete ; very little of one object is seen at one time.

Last Suboctile.—I am very rarely able to recall any object whatever with any sort of distinctness. Very occasionally an object or image will recall itself, but even then it is more like a generalised image than an individual one. I seem to be almost destitute of visualising power as under control.

Lowest.—My powers are zero. To my consciousness there is almost no association of memory with objective visual impressions. I recollect the table, but do not see it.

I next proceed to colour, as specified in the third of my questions, and annex a selection from the returns classified on the same principle as in the preceding paragraph.

COLOUR REPRESENTATION.

Highest.—Perfectly distinct, bright, and natural.

First Suboctile.—White cloth, blue china, argand coffee-pot, buff stand with sienna drawing, toast—all clear.

First Octile.—All details seen perfectly.

First Quartile.—Colours distinct and natural till I begin to puzzle over them.

Middlemost.—Fairly distinct, though not certain that they are accurately recalled.

Last Quartile.—Natural, but very indistinct.

Last Octile.—Faint ; can only recall colours by a special effort for each.

Last Suboctile.—Power is nil.

Lowest.—Power is nil.

It may seem surprising that one out of every sixteen persons who are accustomed to use accurate expressions should speak of their mental imagery as perfectly clear and bright ;

but it is so, and many details are added in various returns emphasising the assertion. One of the commonest of these is to the effect, "If I could draw, I am sure I could draw perfectly from my mental image." That some artists, such as Blake, have really done so is beyond dispute, but I have little doubt that there is an unconscious exaggeration in these returns. My reason for saying so is that I have also returns from artists, who say as follows: "My imagery is so clear, that if I had been unable to draw I should have unhesitatingly said that I could draw from it." A foremost painter of the present day has used that expression. He finds deficiencies and gaps when he tries to draw from his mental vision. There is perhaps some analogy between these images and those of "faces in the fire." One may often fancy an exceedingly well-marked face or other object in the burning coals, but probably everybody will find, as I have done, that it is impossible to draw it, for as soon as its outlines are seriously studied, the fancy flies away.

Mr. Flinders Petrie, a contributor of interesting experiments on kindred subjects to *Nature*, informs me that he habitually works out sums by aid of an imaginary sliding rule, which he sets in the desired way and reads off mentally. He does not usually visualise the whole rule, but only that part of it with which he is at the moment concerned (see Plate II. Fig. 34, where, however, the artist has not put in the divisions very correctly). I think this is one of the most striking cases of accurate visualising power it is possible to imagine.

I have a few returns from chess-players who play games blindfolded; but the powers of such men to visualise the separate boards with different sets of men on the different boards, some ivory, some wood, and so forth, are well known, and I need not repeat them. I will rather give the following extract from an article in the *Pall Mall Gazette*, 27th June 1882, on the recent chess tournament at Vienna :—

"The modern feats of blindfold play (without sight of board) greatly surpass those of twenty years ago. Paul Morphy, the American, was the first who made an especial study of this kind of display, playing some seven or eight games blindfold and simultaneously against various inferior opponents, and making lucrative exhibitions in this way. His abilities in this line created a scare among other rivals who had not practised this test of memory. Since his day many chess-players who are gifted with strong and clear memory and power of picturing to the mind the ideal board and men, have carried this branch of exhibition play far beyond Morphy's pitch ; and, contemporaneously with this development, it has become acknowledged that skill in blindfold play is not an absolute test of similarly relative powers over the board : *e.g.* Blackburne and Zukertort can play as many as sixteen, or even twenty, blindfold games at a time, and win about 80 per cent of them at least. Steinitz, who beats them both in match play, does not essay more than six blindfold at a time. Mason does not, to our knowledge, make any *spécialité* at all of this sort."

I have many cases of persons mentally reading off scores when playing the pianoforte, or manuscript when they are making speeches. One statesman has assured me that a certain hesitation in utterance which he has at times, is due to his being plagued by the image of his manuscript speech with its original erasures and corrections. He cannot lay the ghost, and he puzzles in trying to decipher it.

Some few persons see mentally in print every word that is uttered ; they attend to the visual equivalent and not to the sound of the words, and they read them off usually as from a long imaginary strip of paper, such as is unwound from telegraphic instruments. The experiences differ in detail as to size and kind of type, colour of paper, and so forth, but are always the same in the same person.

A well-known frequenter of the Royal Institution tells me that he often craves for an absence of visual perceptions, they are so brilliant and persistent. The Rev. George Henslow speaks of their extreme restlessness ; they oscillate, rotate, and change."

It is a mistake to suppose that sharp sight is accompanied by clear visual memory. I have not a few instances in which the independence of the two faculties is emphatically commented on ; and I have at least one clear case where great interest in outlines and accurate appreciation of straightness, squareness, and the like, is unaccompanied by the power of visualising. Neither does the faculty go with dreaming. I have cases where it is powerful, and at the same time where dreams are rare and faint or altogether absent. One friend tells me that his dreams have not the hundredth part of the vigour of his waking fancies.

The visualising and the identifying powers are by no means necessarily combined. A distinguished writer on metaphysical topics assures me that he is exceptionally quick at recognising a face that he has seen before, but that he cannot call up a mental image of any face with clearness.

Some persons have the power of combining in a single perception more than can be seen at any one moment by the two eyes. It is needless to insist on the fact that all who have two eyes see stereoscopically, and therefore somewhat round a corner. Children, who can focus their eyes on very near objects, must be able to comprise in a single mental image much more than a half of any small object they are examining. Animals such as hares, whose eyes are set more on the side of the head than ours, must be able to perceive at one and the same instant more of a panorama than we can. I find that a few persons can, by what they often describe as a kind of touch-sight, visualise at the same moment all round the image of a solid body. Many can do so nearly, but not altogether round that of a terrestrial globe. An eminent mineralogist assures me that he is able to imagine simultaneously all the sides of a crystal with which he is familiar. I may be allowed to quote a curious faculty of my own in respect to this. It is exercised only occasionally and in dreams, or rather in nightmares, but under those circumstances I am perfectly conscious of embracing an entire sphere in a single perception. It appears to lie within my mental eyeball, and to be viewed centripetally.

This power of comprehension is practically attained in many cases by indirect methods. It is a common feat to take in the whole surroundings of an imagined room with such a rapid mental sweep as to leave some doubt whether it has not been viewed simultaneously. Some persons have the habit of viewing objects as though they were partly transparent; thus, if they so dispose a globe in their imagination as to see both its north and south poles at the same time, they will not be able to see its equatorial parts. They can also perceive all the rooms of an imaginary house by a single mental glance, the walls and floors being as if made of glass. A fourth class of persons have the habit of recalling scenes, not from the point of view whence they were observed, but from a distance, and they visualise their own selves as actors on the mental stage. By one or other of these ways, the power of seeing the whole of an object, and not merely one aspect of it, is possessed by many persons.

The place where the image appears to lie, differs much. Most persons see it in an indefinable sort of way, others see it in front of the eye, others at a distance corresponding to reality. There exists a power which is rare naturally, but can, I believe, be acquired without much difficulty, of projecting a mental picture upon a piece of paper, and of holding it fast there, so that it can be outlined with a pencil. To this I shall recur.

Images usually do not become stronger by dwelling on them ; the first idea is commonly the most vigorous, but this is not always the case. Sometimes the mental view of a locality is inseparably connected with the sense of its position as regards the points of the compass, real or imaginary. I have received full and curious descriptions from very different sources of this strong geographical tendency, and in one or two cases I have reason to think it allied to a considerable faculty of geographical comprehension.

The power of visualising is higher in the female sex than in the male, and is somewhat, but not much, higher in public schoolboys than in men. After maturity is reached, the further advance of age does not seem to dim the faculty, but rather the reverse, judging from numerous statements to that effect ; but advancing years are sometimes accompanied by a growing habit of hard abstract thinking, and in these cases—not uncommon among those whom I have questioned—the faculty undoubtedly becomes impaired. There is reason to believe that it is very high in some young children, who seem to spend years of difficulty in distinguishing between the subjective and objective world. Language and book-learning certainly tend to dull it.

The visualising faculty is a natural gift, and, like all natural gifts, has a tendency to be inherited. In this faculty the tendency to inheritance is exceptionally strong, as I have abundant evidence to prove, especially in respect to certain rather rare peculiarities, of which I shall speak in the next chapter, and which, when they exist at all, are usually found among two, three, or more brothers and sisters, parents, children, uncles and aunts, and cousins.

Since families differ so much in respect to this gift, we may suppose that races would also differ, and there can be no doubt that such is the case. I hardly like to refer to civilised nations, because their natural faculties are too much modified by education to allow of their being appraised in an off-hand fashion. I may, however, speak of the French, who appear to possess the visualising faculty in a high degree. The peculiar ability they show in prearranging ceremonials and *fêtes* of all kinds, and their undoubted genius for tactics and strategy, show that they are able to foresee effects with unusual clearness. Their ingenuity in all technical contrivances is an additional testimony in the same direction, and so is their singular clearness of expression. Their phrase, "figurez-vous," or "picture to yourself," seems to express their dominant mode of perception. Our equivalent of "imagine" is ambiguous.

It is among uncivilised races that natural differences in the visualising faculty are most conspicuous. Many of them make carvings and rude illustrations, but only a few have the gift of carrying a picture in their mind's eye, judging by the completeness and firmness of their designs, which show no trace of having been elaborated in that step-by-step manner which is characteristic of draughtsmen who are not natural artists.

Among the races who are thus gifted are the commonly despised, but, as I confidently maintain from personal knowledge of them, the much underrated Bushmen of South Africa. They are no doubt deficient in the natural instincts necessary to civilisation, for they detest a regular life, they are inveterate thieves, and are incapable of withstanding the temptation of strong drink. On the other hand, they have few superiors among barbarians in the ingenious methods by which they supply the wants of a difficult existence, and in the effectiveness and nattiness of their accoutrements. One of their habits is to draw pictures on the walls of caves of men and animals, and to colour them with ochre. These drawings were once numerous, but they have been sadly destroyed by advancing colonisation, and few of them, and indeed few wild Bushmen, now exist. Fortunately a large and valuable collection of facsimiles of Bushman art was made before it became too late by Mr. Stow, of the Cape Colony, who has very lately sent some specimens of them to this country, in the hope that means might be found for the publication of the entire series. Among the many pictures of animals in each of the large sheets full of them, I was particularly struck with one of an eland as giving a just idea of the precision and purity of their best work. Others, again, were exhibited last summer at the Anthropological Institute by Mr. Hutchinson.

The method by which the Bushmen draw is described in the following extract from a letter written to me by Dr. Mann, the well-known authority on South African matters of science. The boy to whom he refers belonged to a wild tribe living in caves in the Drakenberg, who plundered outlying farms, and were pursued by the neighbouring colonists. He was wounded and captured, then sent to hospital, and subsequently taken into service. He was under Dr. Mann's observation in the year 1860, and has recently died, to the great regret of his employer, Mr. Proudfoot, to whom he became a valuable servant.

Dr. Mann writes as follows :—

"This lad was very skilful in the proverbial Bushman art of drawing animal figures, and upon several occasions I induced him to show me how this was managed among his people. He invariably began by jotting down upon paper or on a slate a number of isolated dots which presented no connection or trace of outline of any kind to the uninitiated eye, but looked like the stars scattered promiscuously in the sky. Having with much deliberation satisfied himself of the sufficiency of these dots, he forthwith began to run a free bold line from one to the other, and as he did so the form of an animal—horse, buffalo, elephant, or some kind of antelope—gradually developed itself. This was invariably done with a free hand, and with such unerring accuracy of touch, that no correction of a line was at any time attempted. I understood from the lad that this was the plan which was invariably pursued by his kindred in making their clever pictures."

It is impossible, I think, for a drawing to be made on this method unless the artist had a clear image in his mind's eye of what he was about to draw, and was able, in some degree, to project it on the paper or slate.

Other living races have the gift of drawing, but none more so than the Eskimo. I will therefore speak of these and not of the Australian and Tasmanian pictures, nor of the still ruder performances of the old inhabitants of Guiana, nor of those of some North American tribes, as the Iroquois. The Eskimos are geographers by instinct, and appear to see vast tracts of country mapped out in their heads. From the multitude of illustrations of their map-drawing powers, I may mention one of those included in the journals of Captain Hall, at p. 224, which were published in 1879 by the United States Government, under the editorship of Professor J. E. Nourse. It is the facsimile of a chart drawn by an Eskimo who was a thorough barbarian in the accepted sense of the word; that is to say, he spoke no language besides his own uncouth tongue, he was wholly uneducated according to our modern ideas, and he lived in what we should call a savage fashion. This man drew from memory a chart of the region over which he had at one time or another gone in his canoe. It extended from Pond's Bay, in lat. 73°, to Fort Churchill, in lat. 58°44′, over a distance in a straight line of more than 960 nautical, or 1100 English miles, the coast being so indented by arms of the sea that its length is six times as great. On comparing this rough Eskimo outline with the Admiralty chart of 1870, their accordance is remarkable. I have seen many MS. route maps made by travellers a few years since, when the scientific exploration of the world was much less advanced than it is now, and I can confidently say that I have never known of any traveller, white or brown, civilised or uncivilised, in Africa, Asia, or Australia, who, being unprovided with surveying instruments, and trusting to his memory alone, has produced a chart comparable in extent and accuracy to that of this barbarous Eskimo. The aptitude of the Eskimos to draw, is abundantly shown by the numerous illustrations in Rink's work, all of which were made by self-taught men, and are thoroughly realistic.

So much for the wild races of the present day; but even the Eskimo are equalled in their power of drawing by the

men of old times. In ages so far gone by, that the interval
that separates them from our own may be measured in
perhaps hundreds of thousands of years, when Europe was
mostly icebound, a race who, in the opinion of all anthropo-
logists, was closely allied to the modern Eskimo, lived in
caves in the more habitable places. Many broken relics of
that race have been found; some few of these are of
bone engraved with flints or carved into figures, and
among these are representations of the mammoth, elk, and
reindeer, which, if made by an English labourer with the
much better implements at his command, would certainly
attract local attention and lead to his being properly
educated, and in much likelihood to his becoming a
considerable artist if he had intellectual powers to match.

It is not at all improbable that these prehistoric men
had the same geographical instincts as the modern Eskimo,
whom they closely resemble in every known respect. If so,
it is perfectly possible that scraps of charts scratched on
bone or stone, of prehistoric Europe, when the distribution
of land, sea, and ice was very different to what it is now,
may still exist, buried underground, and may reward the
zeal of some future cave explorer.

There is abundant evidence that the visualising faculty
admits of being developed by education. The testimony
on which I would lay especial stress is derived from the
published experiences of M. Lecoq de Boisbaudran, late
director of the École Nationale de Dessein, in Paris, which
are related in his *Education de la Mémoire Pittoresque*.[1]
He trained his pupils with extraordinary success, beginning
with the simplest figures. They were made to study the
models thoroughly before they tried to draw them from
memory. One favourite expedient was to associate the
sight memory with the muscular memory, by making his
pupils follow at a distance the outlines of the figures with
a pencil held in their hands. After three or four months'
practice, their visual memory became greatly strengthened.
They had no difficulty in summoning images at will, in
holding them steady, and in drawing them. Their copies

[1] Republished in an 8vo, entitled *Enseignment Artistique*. Morel
et Cie. Paris, 1879.

were executed with marvellous fidelity, as attested by a commission of the Institute, appointed in 1852 to inquire into the matter, of which the eminent painter Horace Vernet was a member. The present Slade Professor of Fine Arts at University College, M. Légros, was a pupil of M. de Boisbaudran. He has expressed to me his indebtedness to the system, and he has assured me of his own success in teaching others in a somewhat similar way.

Colonel Moncrieff informs me that, when wintering in 1877 near Fort Garry in North America, young Indians occasionally came to his quarters, and that he found them much interested in any pictures or prints that were put before them. On one of these occasions he saw an Indian tracing the outline of a print from the *Illustrated News* very carefully with the point of his knife. The reason he gave for this odd manœuvre was, that he would remember the better how to carve it when he returned home.

I could mention instances within my own experience in which the visualising faculty has become strengthened by practice ; notably one of an eminent electrical engineer, who had the power of recalling form with unusual precision, but not colour. A few weeks after he had replied to my questions, he told me that my inquiries had induced him to practise his colour memory, and that he had done so with such success that he was become quite an adept at it, and that the newly-acquired power was a source of much pleasure to him.

A useful faculty, easily developed by practice, is that of retaining a retinal picture. A scene is flashed upon the eye ; the memory of it persists, and details, which escaped observation during the brief time when it was actually seen, may be analysed and studied at leisure in the subsequent vision.

The memories we should aim at acquiring are, however, such as are based on a thorough understanding of the objects observed. In no case is this more surely effected than in the processes of mechanical drawing, where the intended structure has to be portrayed so exactly in plan, elevation, side view, and sections, that the workman has simply to copy the drawing in metal, wood, or stone, as the case may be. It is undoubtedly the fact that mechanicians, engineers, and architects usually possess the faculty of seeing mental images with remarkable clearness and precision.

A few dots like those used by the Bushmen give great assistance in creating an imaginary picture, as proved by our general habit of working out ideas by the help of marks and rude lines. The use of dolls by children also testifies to the value of an objective support in the construction of mental images. The doll serves as a kind of skeleton for the child to clothe with fantastic attributes, and the less individuality the doll has, the more it is appreciated by the child, who can the better utilise it as a lay figure in many different characters. The chief art of strengthening visual, as well as every other form of memory, lies in multiplying associations; the healthiest memory being that in which all the associations are logical, and toward which all the senses concur in their due proportions. It is wonderful how much the vividness of a recollection is increased when two or more lines of association are simultaneously excited. Thus the inside of a known house is much better visualised when we are looking at its outside than when we are away from it, and some chess-players have told me that it is easier for them to play a game from memory when they have a blank board before them than when they have not.

There is an absence of flexibility in the mental imagery of most persons. They find that the first image they have acquired of any scene is apt to hold its place tenaciously in spite of subsequent need of correction. They find a difficulty in shifting their mental view of an object, and examining it at pleasure in different positions. If they see an object equally often in many positions the memories combine and confuse one another, forming a "composite" blur, which they cannot dissect into its components. They are less able to visualise the features of intimate friends than those of persons of whom they have caught only a single glance. Many such persons have expressed to me their grief at finding themselves powerless to recall the looks of dear relations whom they had lost, while they had no difficulty in recollecting faces that were uninteresting to them.

Others have a complete mastery over their mental images. They can call up the figure of a friend and make it sit on a chair or stand up at will; they can make it turn round and attitudinise in any way, as by mounting it on a bicycle or compelling it to perform gymnastic feats on a trapeze. They are able to build up elaborate geometric structures bit by bit in their mind's eye, and add, subtract, or alter at will and at leisure.   This free action of a vivid visualising faculty is of much importance in connection with the higher processes of generalised thought, though it is commonly put to no such purpose, as may be easily explained by an example.   Suppose a person suddenly to accost another with the following words:—" I want to tell you about a boat."   What is the idea that the word "boat" would be likely to call up?   I tried the experiment with this result.   One person, a young lady, said that she immediately saw the image of a rather large boat pushing off from the shore, and that it was full of ladies and gentlemen, the ladies being dressed in white and blue. It is obvious that a tendency to give so specific an interpretation to a general word is absolutely opposed to philosophic thought.   Another person, who was accustomed to philosophise, said that the word "boat" had aroused no definite image, because he had purposely held his mind in suspense. He had exerted himself not to lapse into any one of the special ideas that he felt the word boat was ready to call up, such as a skiff, wherry, barge, launch, punt, or dingy.   Much more did he refuse to think of any one of these with any particular freight or from any particular point of view.   A habit of suppressing mental imagery must therefore characterise men who deal much with abstract ideas; and as the power of dealing easily and firmly with these ideas is the surest criterion of a high order of intellect, we should expect that the visualising faculty would be starved by disuse among philosophers, and this is precisely what I found on inquiry to be the case.

But there is no reason why it should be so, if the faculty is free in its action, and not tied to reproduce hard and persistent forms; it may then produce generalised pictures out of its past experiences quite automatically.   It has no difficulty in reducing images to the same scale, owing to our constant practice in watching objects as they approach or recede, and consequently grow or diminish in apparent size. It readily shifts images to any desired point of the field of view, owing to our habit of looking at bodies in motion to the right or left, upward or downward. It selects images that present the same aspect, either by a simple act of memory or by a feat of imagination that forces them into the desired position, and it has little or no difficulty in reversing them from right to left, as if seen in a looking-glass. In illustration of these generalised mental images, let us recur to the boat, and suppose the speaker to continue as follows:—"The boat was a four-oared racing-boat, it was passing quickly to the left just in front of me, and the men were bending forward to take a fresh stroke." Now at this point of the story the listener ought to have a picture well before his eye. It ought to have the distinctness of a real four-oar going to the left, at the moment when many of its details still remained unheeded, such as the dresses of the men and their individual features. It would be the generic image of a four-oar formed by the combination into a single picture of a great many sight memories of those boats.

In the highest minds a descriptive word is sufficient to evoke crowds of shadowy associations, each striving to manifest itself. When they differ so much from one another as to be unfitted for combination into a single idea, there will be a conflict, each being prevented by the rest from obtaining sole possession of the field of consciousness. There could, therefore, be no definite imagery so long as the aggregate of all the pictures that the word suggested of objects presenting similar aspects, reduced to the same size, and accurately superposed, resulted in a blur; but a picture would gradually evolve as qualifications were added to the word, and it would attain to the distinctness and vividness of a generic image long before the word had been so restricted as to be individualised. If the intellect be slow, though correct in its operations, the associations will be few, and the generalised image based on insufficient data. If the visualising power be faint, the generalised image will be indistinct.

I cannot discover any closer relation between high visualising power and the intellectual faculties than between verbal memory and those same faculties. That it must afford immense help in some professions stands to reason, but in ordinary social life the possession of a high visualising power, as of a high verbal memory, may pass quite unobserved.

I have to the last failed in anticipating the character of the answers that my friends would give to my inquiries, judging from my previous knowledge of them; though I am bound to say that, having received their answers, I could usually persuade myself that they were justified by my recollections of their previous sayings and conduct generally.

The faculty is undoubtedly useful in a high degree to inventive mechanicians, and the great majority of those whom I have questioned have spoken of their powers as very considerable. They invent their machines as they walk, and see them in height, breadth, and depth as real objects, and they can also see them in action. In fact, a periodic action of any kind appears to be easily recalled. But the powers of other men are considerably less; thus an engineer officer who has himself great power of visual memory, and who has superintended the mathematical education of cadets, doubts if one in ten can visualise an object in three dimensions. I should have thought the faculty would be common among geometricians, but many of the highest seem able somehow to get on without much of it. There is a curious dictum of Napoleon I. quoted in Hume's *Précis of Modern Tactics*, p. 15, of which I can neither find the original authority nor do I fully understand the meaning. He is reported to have said that "there are some who, from some physical or moral peculiarity of character, form a picture (*tableau*) of everything. No matter what knowledge, intellect, courage, or good qualities they may have, these men are unfit to command." It is possible that "tableau" should be construed rather in the sense of a pictorial composition, which, like an epigrammatic sentence, may be very complete and effective, but not altogether true.

There can, however, be no doubt as to the utility of the visualising faculty when it is duly subordinated to the higher intellectual operations. A visual image is the most perfect form of mental representation wherever the shape, position, and relations of objects in space are concerned. It is of importance in every handicraft and profession where design is required. The best workmen are those who visualise the whole of what they propose to do, before they take a tool in their hands. The village smith and the carpenter who are employed on odd jobs employ it no less for their work than the mechanician, the engineer, and the architect. The lady's maid who arranges a new dress requires it for the same reason as the decorator employed on a palace, or the agent who lays out great estates. Strategists, artists of all denominations, physicists who contrive new experiments, and in short all who do not follow routine, have need of it. The pleasure its use can afford is immense. I have many correspondents who say that the delight of recalling beautiful scenery and great works of art is the highest that they know; they carry whole picture galleries in their minds. Our bookish and wordy education tends to repress this valuable gift of nature. A faculty that is of importance in all technical and artistic occupations, that gives accuracy to our perceptions, and justness to our generalisations, is starved by lazy disuse, instead of being cultivated judiciously in such a way as will on the whole bring the best return. I believe that a serious study of the best method of developing and utilising this faculty, without prejudice to the practice of abstract thought in symbols, is one of the many pressing desiderata in the yet unformed science of education.

Number-Forms.

Persons who are imaginative almost invariably think of *numerals* in some form of visual imagery. If the idea of *six* occurs to them, the word " six " does not sound in their mental ear, but the figure 6 in a written or printed form rises before their mental eye. The clearness of the images of numerals, and the number of them that can be mentally viewed at the same time, differs greatly in different persons. The most common case is to see only two or three figures at once, and in a position too vague to admit of definition. There are a few persons in whom the visualising faculty is so low that they can mentally see neither numerals nor anything else; and again there are a few in whom it is so high as to give rise to hallucinations. Those who are able to visualise a numeral with a distinctness comparable to reality, and to behold it as if it were before their eyes, and not in some sort of dreamland, will define the direction in which it seems to lie, and the distance at which it appears to be. If they were looking at a ship on the horizon at the moment that the figure 6 happened to present itself to their minds, they could say whether the image lay to the left or right of the ship, and whether it was above or below the line of the horizon ; they could always point to a definite spot in space, and say with more or less precision that that was the direction in which the image of the figure they were thinking of, first appeared.

Now the strange psychological fact to which I desire to draw attention, is that among persons who visualise figures clearly there are many who notice that the image of the same figure invariably makes its first appearance in the same direction, and at the same distance. Such a person would always see the figure when it first appeared to him at (we may suppose) one point of the compass to the left of the line between his eye and the ship, at the level of the horizon, and at twenty feet distance. Again, we may suppose that he would see the figure 7 invariably half a point to the left of the ship, at an altitude equal to the sun's diameter above the horizon, and at thirty feet distance ; similarly for all the other figures. Consequently, when he thinks of the series of numerals 1, 2, 3, 4, etc., they show themselves in a definite pattern that always occupies an identical position in his field of view with respect to the direction in which he is looking.

Those who do not see figures with the same objectivity, use nevertheless the same expressions with reference to their mental field of view. They can draw what they see in a manner fairly satisfactory to themselves, but they do not locate it so strictly in reference to their axis of sight and to the horizontal plane that passes through it. It is with them as in dreams, the imagery is before and around, but the eyes during sleep are turned inwards and upwards.

The pattern or " Form " in which the numerals are seen is by no means the same in different persons, but assumes the most grotesque variety of shapes, which run in all sorts of angles, bends, curves, and zigzags as represented in the various illustrations to this chapter. The drawings, however, fail in giving the idea of their apparent size to those who see them ; they usually occupy a wider range than the mental eye can take in at a single glance, and compel it to wander. Sometimes they are nearly panoramic.

These Forms have for the most part certain characteristics in common. They are stated in all cases to have been in existence, so far as the earlier numbers in the Form are concerned, as long back as the memory extends; they come into view quite independently of the will, and their shape and position, at all events in the mental field of view, is nearly invariable. They have other points in common to which I shall shortly draw attention, but first I will endeavour to remove all doubt as to the authenticity and trustworthiness of these statements.

I see no " Form " myself, and first ascertained that such a thing existed through a letter from Mr. G. Bidder, Q.C., in which he described his own case as a very curious peculiarity. I was at the time making inquiries about the strength of the visualising faculty in different persons, and among the numerous replies that reached me I soon collected ten or twelve other cases in which the writers spoke of their seeing numerals in definite forms. Though the information came from independent sources, the expressions used were so closely alike that they strongly corroborated one another. Of course I eagerly followed up the inquiry, and when I had collected enough material to justify publication, I wrote an account which appeared in *Nature* on 15th January 1880, with several illustrations. This has led to a wide correspondence and to a much-increased store of information, which enables me to arrive at the following conclusions. The answers I received whenever I have pushed my questions, have been straightforward and precise. I have not unfrequently procured a second sketch of the Form even after more than two years' interval, and found it to agree closely with the first one. I have also questioned many of my own friends in general terms as to whether they visualise numbers in any particular way. The large majority are unable to do so. But every now and then I meet with persons who possess the faculty, and I have become familiar with the quick look of intelligence with which they receive my question. It is as though some chord had been struck which had not been struck before, and the verbal answers they give me are precisely of the same type as those written ones of which I have now so many. I cannot doubt of the authenticity of independent statements which closely confirm one another, nor of the general accuracy of the accompanying sketches, because I find now that my collection is large enough for classification, that they might be arranged in an approximately continuous series. I am often told that the peculiarity is common to the speaker and to some near relative, and that they had found such to be the case by accident. I have the strongest evidence of its hereditary character after allowing, and over-allowing, for all conceivable influences of education and family tradition.

Last of all, I took advantage of the opportunity afforded by a meeting of the Anthropological Institute to read a memoir there on the subject, and to bring with me many gentlemen well known in the scientific world, who have this habit of seeing numerals in Forms, and whose diagrams were suspended on the walls. Amongst them are Mr. G. Bidder, Q.C., the Rev. Mr. G. Henslow, the botanist; Prof. Schuster, F.R.S., the physicist; Mr. Roget, Mr. Woodd Smith, and Colonel Yule, C.B., the geographer. These diagrams are given in Plate I. Figs. 20–24. I wished that some of my foreign correspondents could also have been present, such as M. Antoine d'Abbadie, the well-known French traveller and Membre de l'Institut, and Baron v. Osten Sacken, the Russian diplomatist and entomologist, for they had given and procured me much information.

I feel sure that I have now said enough to remove doubts as to the authenticity of my data. Their trustworthiness will, I trust, be still more apparent as I proceed; it has been abundantly manifest to myself from the internal evidences in a large mass of correspondence, to which I can unfortunately do no adequate justice in a brief memoir. It remains to treat the data in the same way as any other scientific facts and to extract as much meaning from them as possible.

The peculiarity in question is found, speaking very roughly, in about 1 out of every 30 adult males or 15 females. It consists in the sudden and automatic appearance of a vivid and invariable " Form " in the mental field of view, whenever a numeral is thought of, in which each numeral has its own definite place. This Form may consist of a mere line of any shape, of a peculiarly arranged row or rows of figures, or of a shaded space.

I give woodcuts of representative specimens of these Forms, and very brief descriptions of them extracted from the letters of my correspondents. Sixty-three other diagrams on a smaller scale will be found in Plates I., II. and III., and two more which are coloured are given in Plate IV.

D. A. " From the very first I have seen numerals up to nearly 200, range themselves always in a particular manner, and

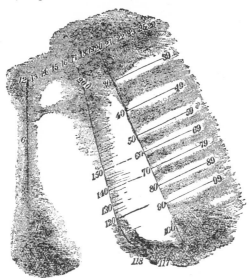

in thinking of a number it always takes its place in the figure. The more attention I give to the properties of numbers and their interpretations, the less I am troubled with this clumsy framework for them, but it is indelible in my mind's eye even when for a long time less consciously so. The higher numbers are to me quite abstract and unconnected with a shape. This rough and untidy [1] production is the best I can do towards repre-

[1] The engraver took much pains to interpret the meaning of the rather faint but carefully made drawing, by strengthening some of the shades. The result was very very satisfactory, judging from the author's own view of it, which is as follows:—" Certainly if the engraver has been as successful with all the other representations as with that of my shape and its accompaniments, your article must be entirely correct."

senting what I see.   There was a little difficulty in the performance, because it is only by catching oneself at unawares, so to
speak, that one is quite sure that what one sees is not affected
by temporary imagination.   But it does not seem much like,
chiefly because the mental picture never seems *on* the flat but
in a thick, dark gray atmosphere deepening in certain parts,
especially where 1 emerges, and about 20.   How I get from
100 to 120 I hardly know, though if I could require these figures
a few times without thinking of them on purpose, I should soon
notice.   About 200 I lose all framework.   I do not see the
actual figures very distinctly, but what there is of them is
distinguished from the dark by a thin whitish tracing.   It is the
place they take and the shape they make collectively which is
invariable.   Nothing more definitely takes its place than a
person's age.   The person is usually there so long as his age is
in mind."

T. M.   "The representation I carry in my mind of the
numerical series is quite distinct to me, so much so that I
cannot think of any number but I at once see it (as it were) in
its peculiar place in the diagram.   My remembrance of dates is
also nearly entirely dependent on a clear mental vision of their
loci in the diagram.   This, as nearly as I can draw it, is the
following :—

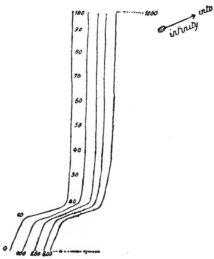

It is only approximately correct (if the term 'correct' be at all applicable). The numbers seem to approach more closely as I ascend from 10 to 20, 30, 40, etc. The lines embracing a hundred numbers also seem to approach as I go on to 400, 500, to 1000. Beyond 1000 I have only the sense of an infinite line in the direction of the arrow, losing itself in darkness towards the millions. Any special number of thousands returns in my mind to its position in the parallel lines from 1 to 1000. The diagram was present in my mind from early childhood; I remember that I learnt the multiplication table by reference to it at the age of seven or eight. I need hardly say that the impression is not that of perfectly straight lines, I have therefore used no ruler in drawing it."

J. S. "The figures are about a quarter of an inch in length, and in ordinary type. They are black on a white ground. The numeral 200 generally takes the place of 100 and obliterates it. There is no light or shade, and the picture is invariable."

In some cases, the mental eye has to travel along the faintly-marked and blank paths of a Form, to the place where the numeral that is wanted is known to reside, and then the figure starts into sight. In other cases all the numerals, as far as 100 or more, are faintly seen at once, but the figure that is wanted grows more vivid than its neighbours; in one of the cases there is, as it were, a chain, and the particular link rises as if an unseen hand had lifted it. The Forms are sometimes variously coloured, occasionally very brilliantly (see Plate IV.). In all of these the definition and illumination vary much in different parts. Usually the Forms fade away into indistinctness after 100; sometimes they come to a dead stop. The higher numbers very rarely fill so large a space in the Forms as the lower ones, and the diminution of space occupied by them is so increasingly rapid that I thought it not impossible they might diminish according to some geometrical law, such as that which governs sensitivity. I took many careful measurements and averaged them, but the result did not justify the supposition.

It is beyond dispute that these forms originate at an early age ; they are subsequently often developed in boyhood and youth so as to include the higher numbers, and, among mathematical students, the negative values.

Nearly all of my correspondents speak with confidence of their Forms having been in existence as far back as they recollect. One states that he knows he possessed it at the age of four ; another, that he learnt his multiplication table by the aid of the elaborate mental diagram he still uses. Not one in ten is able to suggest any clue as to their origin. They cannot be due to anything written or printed, because they do not simulate what is found in ordinary writings or books.

About one-third of the figures are curved to the left, two-thirds to the right ; they run more often upward than downward. They do not commonly lie in a single plane. Sometimes a Form has twists as well as bends, sometimes it is turned upside down, sometimes it plunges into an abyss of immeasurable depth, or it rises and disappears in the sky. My correspondents are often in difficulties when trying to draw them in perspective. One sent me a stereoscopic picture photographed from a wire that had been bent into the proper shape. In one case the Form proceeds at first straightforward, then it makes a backward sweep high above head, and finally recurves into the pocket, of all places ! It is often sloped upwards at a slight inclination from a little below the level of the eye, just as objects on a table would appear to a child whose chin was barely above it.

It may seem strange that children should have such bold conceptions as of curves sweeping loftily upward or downward to immeasurable depths, but I think it may be accounted for by their much larger personal experience of the vertical dimension of space than adults. They are lifted, tossed and swung, but adults pass their lives very much on a level, and only judge of heights by inference from the picture on their retina. Whenever a man first ventures up in a balloon, or is let, like a gatherer of sea-birds' eggs, over the face of a precipice, he is conscious of having acquired a much extended experience of the third dimension of space.

The character of the forms under which historical dates are visualised contrast strongly with the ordinary Number-Forms. They are sometimes copied from the numerical ones, but they are more commonly based both clearly and consciously on the diagrams used in the schoolroom or on some recollected fancy.

The months of the year are usually perceived as ovals, and they as often follow one another in a reverse direction to those of the figures on the clock, as in the same direction. It is a common peculiarity that the months do not occupy equal spaces, but those that are most important to the child extend more widely than the rest. There are many varieties as to the topmost month; it is by no means always January.

The Forms of the letters of the alphabet, when imaged, as they sometimes are, in that way, are equally easy to be accounted for, therefore the ordinary Number-Form is the oldest of all, and consequently the most interesting. I suppose that it first came into existence when the child was learning to count, and was used by him as a natural mnemonic diagram, to which he referred the spoken words "one," "two," "three," etc. Also, that as soon as he began to read, the visual symbol figures supplanted their verbal sounds, and permanently established themselves on the Form. It therefore existed at an earlier date than that at which the child began to learn to read; it represents his mental processes at a time of which no other record remains; it persists in vigorous activity, and offers itself freely to our examination.

The teachers of many schools and colleges, some in America, have kindly questioned their pupils for me; the results are given in the two first columns of Plate I. It appears that the proportion of young people who see numerals in Forms is greater than that of adults. But for the most part their Forms are neither well defined nor complicated. I conclude that when they are too faint to be of service they are gradually neglected, and become wholly forgotten; while if they are vivid and useful, they increase in vividness and definition by the effect of habitual use. Hence, in adults, the two classes of seers and non-seers are rather sharply defined, the connecting link of intermediate cases which is observable in childhood having disappeared.

These Forms are the most remarkable existing instances of what is called "topical" memory, the essence of which appears to lie in the establishment of a more exact system of division of labour in the different parts of the brain, than is usually carried on. Topical aids to memory are of the greatest service to many persons, and teachers of mnemonics make large use of them, as by advising a speaker to mentally associate the corners, etc., of a room with the chief divisions of the speech he is about to deliver. Those who feel the advantage of these aids most strongly are the most likely to cultivate the use of numerical forms. I have read many books on mnemonics, and cannot doubt their utility to some persons; to myself the system is of no avail whatever, but simply a stumbling-block, nevertheless I am well aware that many of my early associations are fanciful and silly.

The question remains, why do the lines of the Forms run in such strange and peculiar ways? the reply is, that different persons have natural fancies for different lines and curves. Their handwriting shows this, for handwriting is by no means solely dependent on the balance of the muscles of the hand, causing such and such strokes to be made with greater facility than others. Handwriting is greatly modified by the fashion of the time. It is in reality a compromise between what the writer most likes to produce, and what he can produce with the greatest ease to himself. I am sure, too, that I can trace a connection between the general look of the handwritings of my various correspondents and the lines of their Forms. If a spider were to visualise numerals, we might expect he would do so in some web-shaped fashion, and a bee in hexagons. The definite domestic architecture of all animals as seen in their nests and holes shows the universal tendency of each species to pursue their work according to certain definite lines and shapes, which are to them instinctive and in no way, we may presume, logical. The same is seen in the groups and formations of flocks of gregarious animals and in the flights of gregarious birds, among which the wedge-shaped phalanx of wild ducks and the huge globe of soaring storks are as remarkable as any.

I used to be much amused during past travels in watching the different lines of search that were pursued by different persons in looking for objects lost on the ground, when the encampment was being broken up. Different persons had decided idiosyncracies, so much so that if their travelling line of sight could have scored a mark on the ground, I think the system of each person would have been as characteristic as his Number-Form.

Children learn their figures to some extent by those on the clock. I cannot, however, trace the influence of the clock on the Forms in more than a few cases. In two of them the clock-face actually appears, in others it has evidently had a strong influence, and in the rest its influence is indicated, but nothing more. I suppose that the complex Roman numerals in the clock do not fit in sufficiently well with the simpler ideas based upon the Arabic ones.

The other traces of the origin of the Forms that appear here and there, are dominoes, cards, counters, an abacus, the fingers, counting by coins, feet and inches (a yellow carpenter's rule appears in one case with 56 in large figures upon it), the country surrounding the child's home, with its hills and dales, objects in the garden (one scientific man sees the old garden walk and the numeral 7 at a tub sunk in the ground where his father filled his watering-pot). Some associations seem connected with the objects spoken of in the doggerel verses by which children are often taught their numbers.

But the paramount influence proceeds from the names of the numerals. Our nomenclature is perfectly barbarous, and that of other civilised nations is not better than ours, and frequently worse, as the French "quatre-vingt dix-huit," or "four score, ten and eight," instead of ninety-eight. We speak of ten, eleven, twelve, thirteen, etc., in defiance of the beautiful system of decimal notation in which we write those numbers. What we see is one-naught, one-one, one-two, etc., and we should pronounce on that principle, with this proviso, that the word for the "one" having to show both the place and the value, should have a sound suggestive of "one" but not identical with it. Let us suppose it to be the letter *o* pronounced short as in "on," then instead of ten, eleven, twelve, thirteen, etc., we might say *on-naught*, *on-one*, *on-two*, *on-three*, etc.

The conflict between the two systems creates a perplexity, to which conclusive testimony is borne by these numerical forms. In most of them there is a marked hitch at the 12, and this repeats itself at the 120. The run of the lines between 1 and 20 is rarely analogous to that between 20 and 100, where it usually first becomes regular. The 'teens frequently occupy a larger space than their due. It is not easy to define in words the variety of traces of the difficulty and annoyance caused by our unscientific nomenclature, that are portrayed vividly, and, so to speak, painfully in these pictures. They are indelible scars that testify to the effort and ingenuity with which a sort of compromise was struggled for and has finally been effected between the verbal and decimal systems. I am sure that this difficulty is more serious and abiding than has been suspected, not only from the persistency of these twists, which would have long since been smoothed away if they did not continue to subserve some useful purpose, but also from experiments on my own mind. I find I can deal mentally with simple sums with much less strain if I audibly conceive the figures as on-naught, on-one, etc., and I can both dictate and write from dictation with much less trouble when that system or some similar one is adopted. I have little doubt that our nomenclature is a serious though unsuspected hindrance to the ready adoption by the public of a decimal system of weights and measures. Three quarters of the Forms bear a duodecimal impress.

I will now give brief explanations of the Number-Forms drawn in Plates I., II., and III., and in the two front figures in Plate IV.

DESCRIPTION OF PLATE I.

Fig. 1 is by Mr. Walter Larden, science-master of Cheltenham College, who sent me a very interesting and elaborate account of his own case, which by itself would make a memoir ; and he has collected other information for me. The Number-Forms of one of his colleagues and of that

PLATE 1.

Examples of Number-Forms.

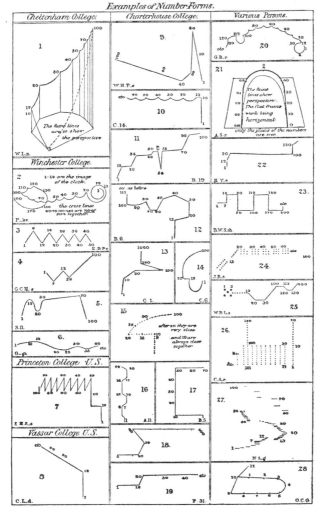

gentleman's sister are given in Figs. 53, 54, Plate III. I extract the following from Mr. Larden's letter—it is all for which I can find space :—

"All numbers are to me as images of figures in general ; I see them in ordinary Arabic type (except in some special cases), and they have definite positions in space (as shown in the Fig.). Beyond 100 I am conscious of coming down a dotted line to the position of 1 again, and of going over the same cycle exactly as before, *e.g.* with 120 in the place of 20, and so on up to 140 or 150. With higher numbers the imagery is less definite ; thus, for 1140, I can only say that there are no new positions, I do not see the entire number in the place of 40 ; but if I think of it as 11 hundred and 40, I see 40 in its place, 11 in its place, and 100 in its place ; the picture is not single though the ideas combine. I seem to stand near 1. I have to turn somewhat to see from 30–40, and more and more to see from 40–100 ; 100 lies high up to my right and behind me. I see no shading nor colour in the figures."

Figs. 2 to 6 are from returns collected for me by the Rev. A. D. Hill, science-master of Winchester College, who sent me replies from 135 boys of an average age of 14–15. He says, speaking of their replies to my numerous questions on visualising generally, that they " represent fairly those who could answer anything ; the boys certainly seemed interested in the subject ; the others, who had no such faculty either attempting and failing, or not finding any response in their minds, took no interest in the inquiry." A very remarkable case of hereditary colour association was sent to me by Mr. Hill, to which I shall refer later. The only five good cases of Number-Forms among the 135 boys are those shown in the Figs. I need only describe Fig. 2. The boy says :— "Numbers, except the first twenty, appear in waves; the two crossing-lines, 60–70, 140–150, never appear at the *same time.* The first twelve are the image of a clock, and 13–20 a continuation of them."

Figs. 7, 8, are sent me by Mr. Henry F. Osborn of Princeton in the United States, who has given cordial assistance in obtaining information as regards visualising generally. These two are the only Forms included in sixty returns that he sent, 34 of which were from Princeton College, and the remaining 26 from Vassar (female) College.

Figs 9–19 and Fig. 28 are from returns communicated by Mr. W. H. Poole, science-master of Charterhouse College, which are very valuable to me as regards visualising power generally. He read my questions before a meeting of about 60 boys, who all consented to reply, and he had several subsequent volunteers. All the answers were short, straightforward, and often amusing. Subsequently the inquiry extended, and I have 168 returns from him in all, containing 12 good Number-Forms, shown in Figs. 9–19, and in Fig. 28. The first Fig. is that of Mr. Poole himself; he says, "The line only represents position; it does not exist in my mind. After 100, I return to my old starting-place, *e.g.* 140 occupies the same position as 40."

The gross statistical result from the schoolboys is as follows:—Total returns, 337: viz. Winchester 135, Princeton 34, Charterhouse 168; the number of these that contained well-defined Number-Forms are 5, 1, and 12 respectively, or total 18—that is, one in twenty. It may justly be said that the masters should not be counted, because it was owing to the accident of their seeing the Number-Forms themselves that they became interested in the inquiry; if this objection be allowed, the proportion would become 16 in 337, or one in twenty-one. Again, some boys who had no visualising faculty at all could make no sense out of the questions, and wholly refrained from answering; this would again diminish the proportion. The shyness in some would help in a statistical return to neutralise the tendency to exaggeration in others, but I do not think there is much room for correction on either head. Neither do I think it requisite to make much allowance for inaccurate answers, as the tone of the replies is simple and straightforward. Those from Princeton, where the students are older and had been specially warned, are remarkable for indications of self-restraint. The result of personal inquiries among adults, quite independent of and prior to these, gave me the proportion of 1 in 30 as a provisional result for adults. This is as well confirmed by the present returns of 1 in 21 among boys and youths as I could have expected.

I have not a sufficient number of returns from girls for useful comparison with the above, though I am much indebted to Miss Lewis for 33 reports, to Miss Cooper of Edgbaston for 10 reports from the female teachers at her school, and to a few other schoolmistresses, such as Miss Stones of Carmarthen, whose returns I have utilised in other ways. The tendency to see Number-Forms is certainly higher in girls than in boys.

Fig. 20 is the Form of Mr. George Bidder, Q.C. ; it is of much interest to myself, because it was, as I have already mentioned, through the receipt of it and an accompanying explanation that my attention was first drawn to the subject. Mr. G. Bidder is son of the late well-known engineer, the famous " calculating boy " of the bygone generation, whose marvellous feats in mental arithmetic were a standing wonder. The faculty is hereditary. Mr. G. Bidder himself has multiplied mentally fifteen figures by another fifteen figures, but with less facility than his father. It has been again transmitted, though in an again reduced degree, to the third generation. He says :—

" One of the most curious peculiarities in my own case is the arrangement of the arithmetical numerals. I have sketched this to the best of my ability. Every number (at least within the first thousand, and afterwards thousands take the place of units) is always thought of by me in its own definite place in the series, where it has, if I may say so, a home and an individuality. I should, however, qualify this by saying that when I am multiplying together two large numbers, my mind is engrossed in the operation, and the idea of locality in the series for the moment sinks out of prominence."

Fig. 21 is that of Prof. Schuster, F.R.S., whose visualising powers are of a very high order, and who has given me valuable information, but want of space compels me to extract very briefly. He says to the effect :—

" The diagram of numerals which I usually see has roughly the shape of a horse-shoe, lying on a slightly inclined plane, with the open end towards me. It always comes into view in front of me, a little to the left, so that the right hand branch of the horse-shoe, at the bottom of which I place 0, is in front of my left eye. When I move my eyes without moving my head, the diagram remains fixed in space and does not follow the movement of my eye. When I move the head the diagram unconsciously follows the movement, but I can, by an effort, keep it fixed in space as before. I can also shift it from one part of the field to the other, and even turn it upside down. I use the diagram as a resting-place for the memory, placing a number on it and finding it again when wanted. A remarkable property of the diagram is a sort of elasticity which enables me to join the two ends of the horse-shoe together when I want to connect 100 with o. The same elasticity causes me to see that part of the diagram on which I fix my attention larger than the rest."

Mr. Schuster makes occasional use of a simpler form of diagram, which is little more than a straight line variously divided, and which I need not describe in detail.

Fig. 22 is by Colonel Yule, C.B.; it is simpler than the others, and he has found it to become sensibly weaker in later years; it is now faint and hard to fix.

Fig. 23. Mr. Woodd Smith :—

"Above 200 the form becomes vague and is soon lost, except that 999 is always in a corner like 99. My own position in regard to it is generally nearly opposite my own age, which is fifty now, at which point I can face either towards 7–12, or towards 12–20, or 20–7, but never (I think) with my back to 12–20."

Fig. 24. Mr. Roget. He writes to the effect that the first twelve are clearly derived from the spots in dominoes. After 100 there is nothing clear but 108. The form is so deeply engraven in his mind that a strong effort of the will was required to substitute any artificial arrangement in its place. His father, the late Dr. Roget (well known for many years as secretary of the Royal Society), had trained him in his childhood to the use of the *memoria technica* of Feinagle, in which each year has its special place in the walls of a particular room, and the rooms of a house represent successive centuries, but he never could locate them in that way. They *would* go to what seemed their natural homes in the arrangement shown in the figure, which had come to him from some unknown source.

The remaining Figs., 25–28, in Plate I., sufficiently

express themselves. The last belongs to one of the Charter-house boys, the others respectively to a musical critic, to a clergyman, and to a gentleman who is, I believe, now a barrister.

Description of Plate II.

Plate II. contains examples of more complicated Forms, which severally require so much minuteness of description that I am in despair of being able to do justice to them separately, and must leave most of them to tell their own story.

Fig. 34 is that of Mr. Flinders Petrie, to which I have already referred (p. 66).

Fig. 37 is by Professor Herbert M'Leod, F.R.S. I will quote his letter almost in full, as it is a very good example :—

" When your first article on visualised numerals appeared in *Nature*, I thought of writing to tell you of my own case, of which I had never previously spoken to any one, and which I never contemplated putting on paper. It becomes now a duty to me to do so, for it is a fourth case of the influence of the clock-face. [In my article I had spoken of only three cases known to me.—F. G.] The enclosed paper will give you a rough notion of the apparent positions of numbers in my mind. That it is due to learning the clock is, I think, proved by my being able to tell the clock certainly before I was four, and probably when little more than three, but my mother cannot tell me the exact date. I had a habit of arranging my spoon and fork on my plate to indicate the positions of the hands, and I well remember being astonished at seeing an old watch of my grandmother's which had ordinary numerals in place of Roman ones. All this happened before I could read, and I have no recollection of learning the numbers unless it was by seeing numbers stencilled on the barrels in my father's brewery.

" When learning the numbers from 12 to 20, they appeared to be vertically above the 12 of the clock, and you will see from the enclosed sketch that the most prominent numbers which I have underlined all occur in the multiplication table. Those doubly underlined are the most prominent [the lithographer has not rendered these correctly.—F. G.], and just now I caught myself doing what I did not anticipate—after doubly underlining some of the numbers, I found that all the multiples of 12 except 84 are so marked. In the sketch I have written in all the numbers up to 30 ; the others are not added merely for want of space ; they

PLATE II.
Examples of Number Forms.

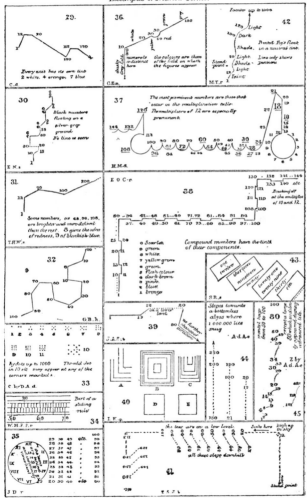

appear in their corresponding positions. You will see that 21 is curiously placed, probably to get a fresh start for the next 10. The loops gradually diminish in size as the numbers rise, and it seems rather curious that the numbers from 100 to 120 resemble in form those from 1 to 20. Beyond 144 the arrangement is less marked, and beyond 200 they entirely vanish, although there is some hazy recollection of a futile attempt to learn the multiplication table up to 20 times 20.

"Neither my mother nor my sister is conscious of any mental arrangement of numerals. I have not found any idea of this kind among any of my colleagues to whom I have spoken on the subject, and several of them have ridiculed the notion, and possibly think me a lunatic for having any such feeling. I was showing the scheme to G., shortly after your first article appeared, on the piece of paper I enclose, and he changed the diagram to a sea-serpent [most amusingly and grotesquely drawn.—F. G.], with the remark, 'If you were a rich man, and I knew I was mentioned in your will, I should destroy that piece of paper, in case it should be brought forward as an evidence of insanity!' I mention this in connection with a paragraph in your article."

Fig. 40 is, I think, the most complicated form I possess. It was communicated to me by Mr. Woodd Smith as that of Miss L. K., a lady who was governess in a family, whom he had closely questioned both with inquiries of his own and by submitting others subsequently sent by myself. It is impossible to convey its full meaning briefly, and I am not sure that I understand much of the principle of it myself. A shows part only (I have not room for more) of the series 2, 3, 5, 7, 10, 11, 13, 14, 17, 18, 19, each as two sides of a square,—that is, larger or smaller according to the magnitude of the number; 1 does not appear anywhere. C similarly shows part of the series (all divisible by 3) of 6, 9, 15, 21, 27, 30, 33, 39, 60, 63, 66, 69, 90, 93, 96. B shows the way in which most numbers divisible by 4 appear. D shows the form of the numbers 17, 18, 19, 21, 22, 23, 25, 26, 27, 29, 41, 42–49, 81–83, 85–87, 89, 101–103, 105–107, and 109. E shows that of 31, 33–35, 37–39. The other numbers are not clear, viz. 50, 51, 53–55, 57–59. Beyond 100 the arrangement becomes hazy, except that the hundreds and thousands go on again in complete, consecutive, and proportional squares indefinitely. The groups of figures are not seen together, but one or other starts up as the number is thought of. The form has no background, and is always seen *in front*. No Arabic or other figures are seen with it. Experiments were made as to the time required to get these images well in the mental view, by reading to the lady a series of numbers as fast as she could visualise them. The first series consisted of twenty numbers of two figures each—thus, 17, 28, 13, 52, etc.; these were gone through on the first trial in 22 seconds, on the second in 16, and on the third in 26. The second series was more varied, containing numbers of one, two, and three figures—thus 121, 117, 345, 187, 13, 6, 25, etc., and these were gone through in three trials in 25, 25, and 22 seconds respectively, forming a general result of 23 seconds for twenty numbers, or $2\frac{1}{3}$ seconds per number. A noticeable feature in this case is the strict accordance of the scale of the image with the magnitude of the number, and the geometric regularity of the figures. Some that I drew, and sent for the lady to see, did not at all satisfy her eye as to their correctness.

I should say that not a few mental calculators work by bulks rather than by numerals; they arrange concrete magnitudes symmetrically in rank and file like battalions, and march these about. I have one case where each number in a Form seems to bear its own *weight*.

Fig. 45 is a curious instance of a French Member of the Institute, communicated to me by M. Antoine d'Abbadie (whose own Number-Form is shown in Fig. 44):—

" He was asked, why he puts 4 in so conspicuous a place ; he replied, 'You see that such a part of my name (which he wishes to withhold) means 4 in the south of France, which is the cradle of my family ; consequently *quatre est ma raison d'être.*' "

Subsequently, in 1880, M. d'Abbadie wrote:—

"I mentioned the case of a philosopher whose, 4, 14, 24, etc., all step out of the rank in his mind's eye. He had a haze in his mind from 60, I believe [it was 50.— F. G.], up to 80 ; but latterly 80 has sprung out, not like the sergeants 4, 14, 24, but like a captain, farther out still, and five or six times as large as the privates 1, 2, 3, 5, 6, etc. 'Were I superstitious,' said he, 'I should conclude that my death would occur in the 80th year of the century.' The growth of 80 was *sudden*, and has remained constant ever since."

This is the only case known to me of a new stage in the development of a Number-Form being suddenly attained.

DESCRIPTION OF PLATE III.

Plate III. is intended to exhibit some instances of heredity. I have no less than twenty-two families in which this curious tendency is hereditary, and there may be many more of which I am still ignorant. I have found it to extend in at least eight of these beyond the near degrees of parent and child, and brother and sister. Considering that the occurrence is so rare as to exist in only about one in every twenty-five or thirty males, these results are very remarkable, and their trustworthiness is increased by the fact that the hereditary tendency is on the whole the strongest in those cases where the Number-Forms are the most defined and elaborate. I give four instances in which the hereditary tendency is found, not only in having a Form at all, but also in some degree in the shape of the Form.

Figs. 46-49 are those of various members of the Henslow family, where the brothers, sisters, and some children of a sister have the peculiarity.

Figs. 53–54 are those of a master of Cheltenham College and his sister.

Figs. 55–56 are those of a father and son; 57 and 58 belong to the same family.

Figs. 59–60 are those of a brother and sister.

The lower half of the Plate explains itself. The last figure of all, Fig. 65, is of interest, because it was drawn for an intelligent little girl of only 11 years old, after she had been closely questioned by the father, and it was accompanied by elaborate coloured illustrations of months and days of the week. I thought this would be a good test case, so I let the matter drop for two years, and then begged the father to question the child casually, and to send me a fresh account. I asked at the same time if any notes had been kept of the previous letter. Nothing could have come out more satisfactorily. No notes had been kept; the subject

'PLATE III.

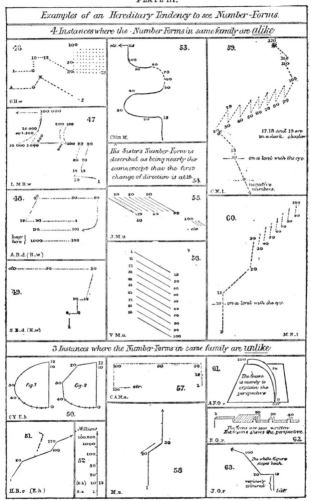

Examples of an Hereditary Tendency to see Number-Forms.

4 Instances where the Number Forms in same family are *alike*

3 Instances where the Number-Forms in same family are *unlike*

had passed out of mind, but the imagery remained the same, with some trifling and very interesting metamorphoses of details.

DESCRIPTION OF PLATE IV.

I can find room in Plate IV. for only two instances of coloured Number-Forms, though others are described in Plate III. Fig. 64 is by Miss Rose G. Kingsley, daughter of the late eminent writer the Rev. Charles Kingsley, and herself an authoress. She says :—

" Up to 30 I see the numbers in clear white ; to 40 in gray ; 40–50 in flaming orange ; 50–60 in green ; 60–70 in dark blue ; 70 I am not sure about ; 80 is reddish, I think ; and 90 is yellow ; but these latter divisions are very indistinct in my mind's eye."

She subsequently writes :—

" I now enclose my diagram ; it is very roughly done, I am afraid, not nearly as well as I should have liked to have done it. My great fear has been that in thinking it over I might be led to write down something more than what I actually see, but I hope I have avoided this."

Fig. 65 is an attempt at reproducing the form sent by Mr. George F. Smythe of Ohio, an American correspondent who has contributed much of interest. He says :—

" To me the numbers from 1 to 20 lie on a level plane, but from 20 they slope up to 100 at an angle of about 25°. Beyond 100 they are generally all on a level, but if for any reason I have to think of the numbers from 100 to 200, or from 200 to 300, etc., then the numbers, between these two hundreds, are arranged just as those from 1 to 100 are. I do not, when thinking of a number, picture to myself the figures which represent it, but I do think instantly of the place which it occupies along the line. Moreover, in the case of numbers from 1 to 20 (and, indistinctly, from 20 up to 28 or 30), I always picture the number—not the figures—as occupying a right-angled parallelogram about twice as long as it is broad. These numbers all lie down flat and extend in a straight line from 1 to 12 over an unpleasant, arid, sandy plain. At 12 the line turns abruptly to the right, passes into a pleasanter region where grass grows, and so continues up to 20. At 20 the line turns to the left, and passes up the before-described incline to 100. This figure will help you in under-standing my ridiculous notions. The asterisk (*) marks the place where I commonly seem to myself to stand and view the line. At times I take other positions, but never any position to the left of the *, nor to the right of the line from 20 upwards. I do not associate colours with numbers, but there is a great difference in the illumination which different numbers receive. If a traveller should start at 1 and walk to 100, he would be in an intolerable glare of light until near 9 or 10. But at 11 he would go into a land of darkness and would have to feel his way. At 12 light breaks in again, a pleasant sunshine, which continues up to 19 or 20, where there is a sort of twilight. From here to 40 the illumination is feeble, but still there is considerable light. At 40 things light up, and until one reaches 56 or 57 there is broad daylight. Indeed the tract from 48 to 50 is almost as bad as that from 1 to 9. Beyond 60 there is a fair amount of light up to about 97. From this point to 100 it is rather cloudy."

In a subsequent letter he adds :—

"I enclose a picture in perspective and colour of my 'form.' I have taken great pains with this, but am far from satisfied with it. I know nothing about drawing, and consequently am unable to put upon the paper just what I see. The faults which I find with the picture are these. The rectangles stand out too distinctly, as something lying on the plane instead of being, as they ought, a part of the plane. The view is taken of necessity from an unnatural stand-point, and some way or other the region 1–12 does not look right. The landscape is altogether too distinct in its features. I rather *know that there is* grass, and that there are trees in the distance, than *see* them. But the grass within a few feet of the line I see distinctly. I cannot make the hill at the right slope down to the plane as it ought. It is too steep. I have had my poor success in indicating my notion of the darkness which overhangs the region of eleven. In reality it is not a cloud at all, but a darkness.

"My sister, a married lady, thirty-eight years of age, sees numerals much as I do, but very indistinctly. She cannot draw a figure which is not by far too distinct."

Most of those who associate colours with numerals do so in a vague way, impossible to convey with truth in a painting. Of the few who see them with more objectivity, many are unable to paint or are unwilling to take the trouble required to match the precise colours of their fancies. A slight error in hue or tint always dissatisfies them with their work.

Before dismissing the subject of numerals, I would call

attention to a few other associations connected with them. They are often personified by children, and characters are assigned to them, it may be on account of the part they play in the multiplication table, or owing to some fanciful association with their appearance or their sound. To the minds of some persons the multiplication table appears dramatised, and any chance group of figures may afford a plot for a tale. I have collated six full and trustworthy accounts, and find a curious dissimilarity in the personifications and preferences; thus the number 3 is described as (1) disliked; (2) a treacherous sneak; (3) a good old friend; (4) delightful and amusing; (5) a female companion to 2; (6) a feeble edition of 9. In one point alone do I find any approach to unanimity, and that is in the respect paid to 12, as in the following examples:—(1) important and influential; (2) good and cautious—so good as to be almost noble; (3) a more beautiful number than 10, from the many multiples that make it up—in other words, its kindly relations to so many small numbers; (4) a great love for 12, a large-hearted motherly person because of the number of little ones that it takes, as it were, under its protection. The decimal system seemed to me treason against this motherly 12.—All this concurs with the importance assigned for other reasons to the number 12 in the Number-Form.

There is no agreement as to the sex of numbers; I myself had absurdly enough fancied that *of course* the even numbers would be taken to be of the male sex, and was surprised to find that they were not. I mention this as an example of the curious way in which our minds may be unconsciously prejudiced by the survival of some forgotten early fancies. I cannot find on inquiring of philologists any indications of different sexes having been assigned in any language to different numbers.

Mr. Hershon has published an analysis of the Talmud, on the odd principle of indexing the various passages according to the number they may happen to contain; thus such a phrase as "there were three men who," etc., would be entered under the number 3. I cannot find any particular preferences given there to especial numbers; even 7 occurs less often than 1, 2, 3, 4, and 10. Their respective frequency being 47, 54, 53, 64, 54, 51 ; 12 occurs only sixteen times. Gamblers have not unfrequently the silliest ideas concerning numbers, their heads being filled with notions about lucky figures and beautiful combinations of them. There is a very amusing chapter in *Rome Contemporaine*, by E. About, in which he speaks of this in connection with the rage for lottery tickets.

COLOUR ASSOCIATIONS.

Numerals are occasionally seen in Arabic or other figures, not disposed in any particular Form, but coloured. An instance of this is represented in Fig. 69 towards the middle part of the column, but as I shall have shortly to enter at length into the colour associations of the author, I will pass over this portion of them, and will quote in preference from the letter of another correspondent.

Baron von Osten Sacken, of whom I have already spoken, writes :—

"The localisation of numerals, peculiar to certain persons, is foreign to me. In my mind's eye the figures appear *in front* of me, within a limited space. My peculiarity, however, consists in the fact that the numerals from 1 to 9 are differently coloured; (1) black, (2) yellow, (3) pale brick red, (4) brown, (5) blackish gray, (6) reddish brown, (7) green, (8) bluish, (9) reddish brown, somewhat like 6. These colours appear very distinctly when I think of these figures separately; in compound figures they become less apparent. But the most remarkable manifestation of these colours appears in my recollections of chronology. When I think of the events of a given century they invariably appear to me on a background coloured like the principal figure in the dates of that century ; thus events of the eighteenth century invariably appear to me on a greenish ground, from the colour of the figure 7. This habit clings to me most tenaciously, and the only hypothesis I can form about its origin is the following :—My tutor, when I was ten to twelve years old, taught me chronology by means of a diagram on which the centuries were represented by squares, subdivided in 100 smaller squares ; the squares representing centuries had *narrow coloured borders;* it may be that in this way the recollection of certain figures became associated with certain colours. I venture this explanation without attaching too much importance to it, because it seems to me that if it was true, my *direct* recollection of those coloured borders would have been stronger than it is ; still, the strong association of my chronology with colour seems to plead in favour of that explanation."

Figs. 66, 67. These two are selected out of a large collection of coloured Forms in which the months of the year are visualised. They will illustrate the gorgeousness of the mental imagery of some favoured persons. Of these Fig. 66 is by the wife of an able London physician, and Fig. 67 is by Mrs. Kempe Welch, whose sister, Miss Bevington, a well-known and thoughtful writer, also sees coloured imagery in connection with dates. This Fig. 67 was one of my test cases, repeated after the lapse of two years, and quite satisfactorily. The first communication was a descriptive account, partly in writing, partly by word of mouth ; the second, on my asking for it, was a picture which agreed perfectly with the description, and explained much that I had not understood at the time. The small size of the Fig. in the Plate makes it impossible to do justice to the picture, which is elaborate and on a large scale, with a perspective of similar hills stretching away to the far distance, and each standing for a separate year. She writes :—

"It is rather difficult to give it fully without making it too definite ; on each side there is a total blank."

The instantaneous association of colour with sound characterises a small percentage of adults, and it appears to be rather common, though in an ill-developed degree, among children. I can here appeal not only to my own collection of facts, but to those of others, for the subject has latterly excited some interest in Germany. The first widely known case was that of the brothers Nussbaumer, published in 1873 by Professor Bruhl of Vienna, of which the English reader will find an account in the last volu ˙e of Lewis's *Problems of Life and Mind* (p. 280). Since then many occasional notices of similar associations have appeared. A pamphlet containing numerous cases was published in Leipsic in 1881 by two Swiss investigators,

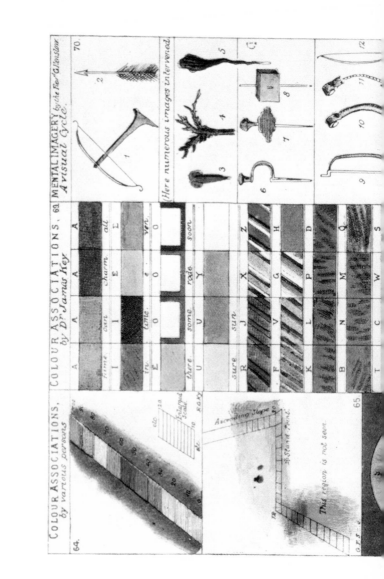

COLOUR ASSOCIATIONS, by various persons

COLOUR ASSOCIATIONS. 69 by Dr Jarius Key

MENTAL IMAGERY by the Rev G Winslow. A visual Cycle.

Here numerous images intervened.

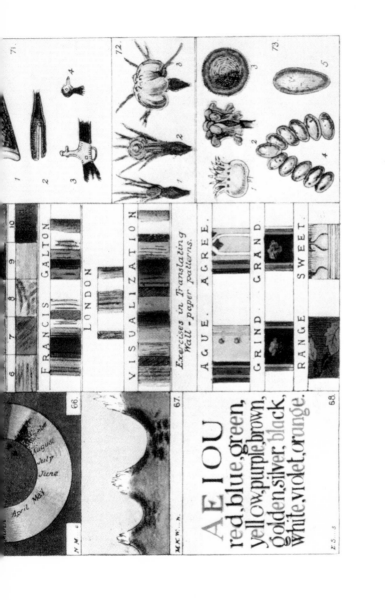

71.

72.

73.

FRANCIS GALTON

LONDON

VISUALIZATION

Exercises in Translating Wall-paper patterns.

AGUE. AGREE.

GRIND. GRAND

RANGE SWEET.

66.

67.

April May June July August

AEIOU
red, blue, green,
yellow, purple, brown,
golden, silver, black,
white, violet, orange.

68.

Messrs. Bleuler and Lehmann.[1] One of the authors had the faculty very strongly, and the other had not; so they worked conjointly with advantage. They carefully tabulated the particulars of sixty-two cases. As my present object is to subordinate details to the general impression that I wish to convey of the peculiarities of different minds, I will simply remark—First, that the persistence of the colour association with sounds is fully as remarkable as that of the Number-Form with numbers. Secondly, that the vowel sounds chiefly evoke them. Thirdly, that the seers are invariably most minute in their description of the precise tint and hue of the colour. They are never satisfied, for instance, with saying "blue," but will take a great deal of trouble to express or to match the particular blue they mean. Fourthly, that no two people agree, or hardly ever do so, as to the colour they associate with the same sound. Lastly, that the tendency is very hereditary. The publications just mentioned absolve me from the necessity of giving many extracts from the numerous letters I have received, but I am particularly anxious to bring the brilliancy of these colour associations more vividly before the reader than is possible by mere description. I have therefore given the elaborately-coloured diagrams in Plate IV., which were copied by the artist directly from the original drawings, and which have been printed by the superimposed impressions of different colours from different lithographic stones. They have been, on the whole, very faithfully executed, and will serve as samples of the most striking cases. Usually the sense of colour is much too vague to enable the seer to reproduce the various tints so definitely as those in this Plate. But this is by no means universally the case.

Fig. 68 is an excellent example of the occasional association of colours with letters. It is by Miss Stones, the head teacher in a high school for girls, who, as I have already mentioned, obtained useful information for me, and has contributed several suggestive remarks of her own. She says:—

[1] Zwangmässige Lichtempfindungen durch Schall und verwandte Erscheinungen, von E. Bleuler und K. Lehmann. Leipsig, Fues' Verlag (R. Reisland), 1881.

"The vowels of the English language always appear to me, when I think of them, as possessing certain colours, of which I enclose a diagram. Consonants, when thought of by themselves, are of a purplish black ; but when I think of a whole word, the colour of the consonants tends towards the colour of the vowels. For example, in the word 'Tuesday,' when I think of each letter separately, the consonants are purplish-black, *u* is a light dove colour, *e* is a pale emerald green, and *a* is yellow ; but when I think of the whole word together, the first part is a light gray-green, and the latter part yellow. Each word is a distinct whole. I have always associated the same colours with the same letters, and no effort will change the colour of one letter, transferring it to another. Thus the word 'red' assumes a light-green tint, while the word 'yellow' is light-green at the beginning and red at the end. Occasionally, when uncertain how a word should be spelt, I have considered what colour it ought to be, and have decided in that way. I believe this has often been a great help to me in spelling, both in English and foreign languages. The colour of the letters is never smeared or blurred in any way. I cannot recall to mind anything that should have first caused me to associate colours with letters, nor can my mother remember any alphabet or reading-book coloured in the way I have described, which I might have used as a child. I do not associate any idea of colour with musical notes at all, nor with any of the other senses.'

She adds :—

" Perhaps you may be interested in the following account from my sister of her visual peculiarities : 'When I think of Wednesday I see a kind of oval flat wash of yellow emerald green ; for Tuesday, a gray sky colour; for Thursday, a brown-red irregular polygon ; and a dull yellow smudge for Friday.'"

The latter quotation is a sample of many that I have ; I give it merely as another instance of hereditary tendency.

I will insert just one description of other coloured letters than those represented in the Plate. It is from Mrs. H., the married sister of a well-known man of science, who writes :—

" I do not know how it is with others, but to me the colours of vowels are so strongly marked that I hardly understand their appearing of a different colour, or, what is nearly as bad, colour-less to any one. To me they are and always have been, as long as I have known them, of the following tints :—

"A, pure white, and like china in texture.

E, red, not transparent; vermilion, with china-white would represent it.

I, light bright yellow; gamboge.

O, black, but transparent; the colour of deep water seen through thick clear ice.

U, purple.

Y, a dingier colour than I.

"The shorter sounds of the vowels are less vivid and pure in colour. Consonants are almost or quite colourless to me, though there is some blackness about M.

"Some association with U in the words blue and purple may account for that colour, and possibly the E in red may have to do with that also; but I feel as if they were independent of suggestions of the kind.

"My first impulse is to say that the association lies solely in the sound of the vowels, in which connection I certainly feel it the most strongly; but then the thought of the distinct redness of such a [printed or written] word as '*great*,' shows me that the relation must be visual as well as aural. The meaning of words is so unavoidably associated with the sight of them, that I think this association rather overrides the primitive impression of the colour of the vowels, and the word '*violet*' reminds me of its proper colour until I look at the word as a mere collection of letters.

"Of my two daughters, one sees the colours quite differently from this (A, blue; E, white; I, black; O, whity-brownish; U, opaque brown). The other is only heterodox on the A and O; A being with her black, and O white. My sister and I never agreed about these colours, and I doubt whether my two brothers feel the chromatic force of the vowels at all."

I give this instance partly on account of the hereditary interest. I could add cases from at least three different families in which the heredity is quite as strongly marked.

Fig. 69 fills the whole of the middle column of Plate IV., and contains specimens from a large series of coloured illustrations, accompanied by many pages of explanation from a correspondent, Dr. James Key of Montagu, Cape Colony. The pictures will tell their own tale sufficiently well. I need only string together a few brief extracts from his letters, as follows:—

"I confess my inability to understand visualised numerals; it is otherwise, however, with regard to colour associations with letters. Ever since childhood these have been distinct and unchanging in my consciousness ; sometimes, although very seldom, I have mentioned them, to the amazement of my teachers and the scorn of my comrades. A is brown. I say it most dogmatically, and nothing will ever have the effect, I am convinced, of making it appear otherwise! I can imagine no explanation of this association. [He goes into much detail as to conceivable reasons connected with his childish life to show that none of these would do.] Shades of brown accompany to my mind the various degrees of openness in pronouncing A. I have never been destitute in all my conscious existence of a conviction that E is a clear, cold, light-gray blue. I remember daubing in colours, when quite a little child, the picture of a jockey, whose shirt received a large share of E, as I said to my-self while daubing it with green. [He thinks that the letter I may possibly be associated with black because it contains no open space, and O with white because it does.] The colour of R has been invariably of a copper colour, in which a swarthy blackness seems to intervene, visually corresponding to the trilled pronunciation of R. This same appearance exists also in J, X, and Z."

The upper row of Fig. 69 shows the various shades of brown, associated with different pronunciations of the letter A, as in " fame," " can," " charm," and " all " respectively. The second, third and fourth rows similarly refer to the various pronunciations of the other vowels. Then follow the letters of the alphabet, grouped according to the character of the appearance they suggest. After these come the numerals. Then I give three lines of words such as they appear to him. The first is my own name, the second is " London," and the third is " Visualisation." Proceeding conversely, Dr. Key collected scraps of various patterns of wall paper, and sent them together with the word that the colour of the several patterns suggested to him. Specimens of these are shown in the three bottom lines of the Fig. I have gone through the whole of them with care, together with his descriptions and reasons, and can quite understand his meaning, and how exceedingly complex and refined these associations are. The patterns are to him like words in poetry, which call up associations that any sub-stituted word of a like dictionary meaning would fail to do. It would not, for example, be possible to print words by the use of counters coloured like those in Fig. 69, because the tint of each influences that of its neighbours. It must be understood that my remarks, though based on Dr. Key's diagrams and statements as on a text, do not depend, by any means, wholly upon them, but on numerous other letters from various quarters to the same effect. At the same time I should say that Dr. Key's elaborate drawings and ample explanations, to which I am totally unable to do justice in a moderate space, are the most full and striking of any I have received. His illustrations are on a large scale, and are ingeniously arranged so as to express his meaning.

Persons who have colour associations are unsparingly critical. To ordinary individuals one of these accounts seems just as wild and lunatic as another, but when the account of one seer is submitted to another seer, who is sure to see the colours in a different way, the latter is scandalised and almost angry at the heresy of the former. I submitted this very account of Dr. Key to a lady, the wife of an ex-governor of one of the most important British possessions, who has vivid colour associations of her own, and who, I had some reason to think, might have personal acquaintance with the locality where Dr. Key lives. She could not comprehend his account at all, his colours were so entirely different to those that she herself saw.

I have now completed as much as I propose to say about the quaint phenomena of Visualised Forms of numbers and of dates, and of coloured associations with letters. I shall not extend my remarks to such subjects as a musician hearing mental music, of which I have many cases, nor to fancies concerning the other senses, as none of these are so noteworthy. I am conscious that the reader may desire even more assurance of the trustworthiness of the accounts I have given than the space now at my disposal admits, or than I could otherwise afford without wearisome iteration of the same tale, by multiplying extracts from my large store of material. I feel, too, that it may seem ungracious to many obliging correspondents not to have made more evident use of what they have sent than my few and brief notices permit. Still their end and mine will have been gained, if these remarks and illustrations succeed in leaving a just impression of the vast variety of mental constitution that exists in the world, and how impossible it is for one man to lay his mind strictly alongside that of another, except in the rare instances of close hereditary resemblance.

Visionaries.

In the course of my inquiries into visual memory, I was greatly struck by the frequency of the replies in which my informants described themselves as subject to "visions." Those of whom I speak were sane and healthy, but were subject notwithstanding to visual presentations, for which they could not account, and which in a few cases reached the level of hallucinations. This unexpected prevalence of a visionary tendency among persons who form a part of ordinary society seems to me suggestive and well worthy of being put on record. The images described by different persons varied greatly in distinctness, some were so faint and evanescent as to appear unworthy of serious notice; others left a deep impression, and others again were so vivid as actually to deceive the judgment. All of these belong to the same category, and it is the assurance of their common origin that affords justification for directing scientific attention to what many may be inclined to contemptuously disregard as the silly vagaries of vacant minds.

The lowest order of phenomena that admit of being classed as visions are the "Number-Forms" to which I have just drawn attention. They are in each case absolutely unchangable, except through a gradual development in complexity. Their diversity is endless, and the Number-Forms of different persons are mutually unintelligible. These strange "visions," for such they must be called, are extremely vivid in some cases, but are almost incredible to the vast majority of mankind, who would set them down as fantastic nonsense; nevertheless, they are familiar parts of the mental furniture of the rest, in whose imaginations they have been unconsciously formed, and where they remain unmodified and unmodifiable by teaching. I have received many touching accounts of their childish experiences from persons who see the Number-Forms, and other curious visions of which I have spoken or shall speak. As is the

case with the colour-blind, so with these seers. They imagined at first that everybody else had the same way of
regarding things as themselves. Then they betrayed their
peculiarities by some chance remark that called forth a stare
of surprise, followed by ridicule and a sharp scolding for
their silliness, so that the poor little things shrank back
into themselves, and never ventured again to allude to their
inner world. I will quote just one of many similar letters
as a sample. I received it, together with much interesting
information, immediately after a lecture I gave to the
British Association at Swansea, in which I had occasion to
speak of the Number-Forms. The writer says:—

"I had no idea for many years that every one did not imagine
numbers in the same positions as those in which they appear to
me. One unfortunate day I spoke of it, and was sharply rebuked
for my absurdity. Being a very sensitive child I felt this acutely,
but nothing ever shook my belief that, absurd or not, I always
saw numbers in this particular way. I began to be ashamed
of what I considered a peculiarity, and to imagine myself, from
this and various other mental beliefs and states, as somewhat
isolated and peculiar. At your lecture the other night, though I
am now over twenty-nine, the memory of my childish misery at
the dread of being peculiar came over me so strongly that I felt I
must thank you for proving that, in this particular at any rate,
my case is most common."

The next sort of vision that flashes unaccountably into
existence is the instant association in some persons of colour
with sound, which was spoken of in the last chapter, and on
which I need not say more now.

A third curious and abiding fantasy of certain persons is
invariably to connect visualised pictures with words, the
same picture to the same word. These are perceived by
many in a vague, fleeting, and variable way, but to a few they
appear strangely vivid and permanent. I have collected
many cases of this peculiarity, and am much indebted to
the authoress, Mrs. Haweis, who sees these pictures, for her
kindness in sketching some of them for me, and for per-
mitting me to use her name in guarantee of their genuineness.
She says:—

"Printed words have always had faces to me; they had
definite expressions, and certain faces made me think of certain

I

words. The words had *no* connection with these except sometimes by accident. The instances I give are few and ridiculous. When I think of the word Beast, it has a face something like a gurgoyle. The word Green has also a gurgoyle face, with the addition of big teeth. The word Blue blinks and looks silly, and turns to the right. The word Attention has the eyes greatly turned to the left. It is difficult to draw them properly because, like Alice's 'Cheshire cat,' which at times became a grin without a cat, these faces have expression without features. The expression of course" [note the *naïve* phrase "of course."— F. G.] "depends greatly on those of the letters, which have likewise their faces and figures. All the little a's turn their eyes to the left, this determines the eyes of Attention. Ant, however, looks a little down. Of course these faces are endless as words are, and it makes my head ache to retain them long enough to draw.'

Some of the figures are very quaint. Thus the interrogation "what?" always excites the idea of a fat man cracking a long whip. They are not the capricious creations of the fancy of the moment, but are the regular concomitants of the words, and have been so as far back as the memory is able to recall.

When in perfect darkness, if the field of view be carefully watched, many persons will find a perpetual series of changes to be going on automatically and wastefully in it. I have much evidence of this. I will give my own experience the first, which is striking to me, because I am very unimpressionable in these matters. I visualise with effort; I am peculiarly inapt to see "after-images," "phosphenes," "light-dust," and other phenomena due to weak sight or sensitiveness; and, again, before I thought of carefully trying, I should have emphatically declared that my field of view in the dark was essentially of a uniform black, subject to an occasional light-purple cloudiness and other small variations. Now, however, after habituating myself to examine it with the same sort of strain that one tries to decipher a signpost in the dark, I have found out that this is by no means the case, but that a kaleidoscopic change of patterns and forms is continually going on, but they are too fugitive and elaborate for me to draw with any approach to truth. I am astonished at their variety, and cannot guess in the remotest degree the cause of them. They disappear out of sight and

memory the instant I begin to think about anything, and it
is curious to me that they should often be so certainly
present and yet be habitually overlooked. If they were
more vivid, the case would be very different, and it is most
easily conceivable that some very slight physiological change,
short of a really morbid character, would enhance their
vividness. My own deficiencies, however, are well supplied
by other drawings in my possession. These are by the Rev.
George Henslow, whose visions are far more vivid than
mine. His experiences are not unlike those of Goethe,
who said, in an often-quoted passage, that whenever he bent
his head and closed his eyes and thought of a rose, a sort of
rosette made its appearance, which would not keep its shape
steady for a moment, but unfolded from within, throwing
out a succession of petals, mostly red but sometimes green,
and that it continued to do so without change in brightness
and without causing him any fatigue so long as he cared to
watch it. Mr. Henslow, when he shuts his eyes and waits,
is sure in a short time to see before him the clear image of
some object or other, but usually not quite natural in its
shape. It then begins to change from one form to another,
in his case also for as long a time as he cares to watch it.
Mr. Henslow has zealously made repeated experiments on
himself, and has drawn what he sees. He has also tried
how far he is able to mould the visions according to his will.
In one case, after much effort, he contrived to bring the
imagery back to its starting-point, and thereby to form what
he terms a "visual cycle." The following account is extracted and condensed from his very interesting letter, and
will explain the illustrations copied from his drawings that
are given in Plate IV.

Fig. 70. The first image that spontaneously presented
itself was a cross-bow (1); this was immediately provided
with an arrow (2), remarkable for its pronounced barb and
superabundance of feathering. Some person, but too
indistinct to recognise much more of him than the hands,
appeared to shoot the arrow from the bow. The single
arrow was then accompanied by a flight of arrows from right
to left, which completely occupied the field of vision. These
changed into falling stars, then into flakes of a heavy snow-
storm; the ground gradually appeared as a sheet of snow

where previously there had been vacant space. Then a well-known rectory, fish-ponds, walls, etc., all covered with snow, came into view most vividly and clearly defined. This somehow suggested another view, impressed on his mind in childhood, of a spring morning, brilliant sun, and a bed of red tulips : the tulips gradually vanished except one, which appeared now to be isolated and to stand in the usual point of sight. It was a single tulip, but became double. The petals then fell off rapidly in a continuous series until there was nothing left but the pistil (3), but (as is almost invariably the case with his objects) that part was greatly exaggerated. The stigmas then changed into three branching brown horns (4); then into a knob (5), while the stalk changed into a stick. A slight bend in it seems to have suggested a centre-bit (6); this passed into a sort of pin passing through a metal plate (7), this again into a lock (8), and afterwards into a nondescript shape (9), distantly suggestive of the original cross-bow. Here Mr. Henslow endeavoured to force his will upon the visions, and to reproduce the cross-bow, but the first attempt was an utter failure. The figure changed into a leather strap with loops (10), but while he still endeavoured to change it into a bow the strap broke, the two ends were separated, but it happened that an imaginary string connected them (11). This was the first concession of his automatic chain of thoughts to his will. By a continued effort the bow came (12), and then no difficulty was felt in converting it into the cross-bow, and thus returning to the starting-point.

Fig. 71. Mr. Henslow writes :—

"Though I can usually summon up any object thought of, it not only is somewhat different from the real thing, but it rapidly changes. The changes are in many cases clearly due to a suggestiveness in the article of something else, but not always so, as in some cases hereafter described. It is not at all necessary to think of any particular object at first, as something is sure to come spontaneously within a minute or two. Some object having once appeared, the automatism of the brain will rapidly induce the series of changes. The images are sometimes very numerous, and very rapid in succession : very frequently of great beauty and highly brilliant. Cut glass (far more elaborate than I am conscious of ever having seen), highly chased gold and silver filigree ornaments ; gold and silver flower-stands, etc. ; elaborate coloured patterns of carpets in brilliant tints are not uncommon.

"Another peculiarity resides in the extreme restlessness of my visual objects. It is often very difficult to keep them still, as well as from changing in character. They will rapidly oscillate or else rotate to a most perplexing degree, and when the characters change at the same time a critical examination is almost impossible. When the process is in full activity, I feel as if I were a mere spectator at a diorama of a very eccentric kind, and was in no way concerned with the getting up of the performance.

"When a succession of images has been passing, I sometimes *determine* to introduce an object, say a watch. Very often it is next to impossible to succeed. There is an evident struggle. The watch, pure and simple, will not come ; but some hybrid structure appears—something round, perhaps—but it lapses into a warming-pan or other unexpected object.

"This practice has brought to my mind very clearly the distinction between at least one form of automatism of the brain and volition ; but the strength of the former is enormous, for the visual objects, when in full career of the change, are *imperative* in their refusal to be interfered with.

"I will now describe the cases illustrated. Fig. 71. I thought of a gun. The *stock* came into view, the metal plate on the end very distinct towards the left (1). The wood was elaborately carved. I cannot recall the pattern. As I scrutinised it, the stock oscillated up and down, and *crumpled up*. The metallic plate sank inwards : and the stock contracted so that it looked not unlike a tuning-fork (2). I gave up the stock and proceeded cautiously to examine the lock. I got it well into view, but no more of the gun. It turned out to be an old-fashioned flint-lock. It immediately began to nod backwards and forwards in a manner suggestive of the beak of a bird pecking. Consequently it forthwith became converted into the head of a bird with a long curved beak, the knob on the lock (3) becoming the head of the bird. I then looked to the right expecting to find the barrel, but the snout of a saw-fish with the tip *distinctly* broken off appeared instead. I had not thought either of a *flint*-lock or of a saw-fish : both came spontaneously.

"Fig. 72. I have several times thought of a rosebud, as Goethe is said to have been able to see one at will, and to observe it expand. The following are some of the results :— The bud appeared unexpectedly a moss rosebud. Its only abnormal appearance was the inordinately elongated sepals (1). I tried to *force* it to expand. It enlarged but only partially opened (2), when all of a sudden it burst open and the petals became reflexed (3).[1]

"Fig. 73. The spontaneous appearance of a poppy capsule (1) dehiscing as usual by 'pores,' but with inordinately long and arching valves over the pores. These valves were eminently suggestive of hooded flowers. Hence they changed to a whorl of *salvias* (2). Each blossom now gyrated rapidly in a vertical plane. Concentrating observation on *one* rotating flower, it became a 'rotating haze,' as the rapid motion rendered the flower totally indistinct. The 'haze' now shaped itself into a circle of moss with a deep funnel-like cavity. This was suggestive of a bird's nest. It became lined with *hair*, but the nest was a *deep*, pointed cavity. A nest was suggestive of eggs. Hence a series appeared (4) ; the two rows meeting in one at the apex appears to have arisen from the *perspective* view of the nest. The eggs all disappeared but one (5), which increased in size ; the bright point of light now shone with great intensity like a star ; then it gradually grew dimmer and dimmer till it disappeared into the usual hazy obscurity into which all [my] visual objects ultimately vanish."

I have a sufficient variety of cases to prove the continuity between all the forms of visualisation, beginning with an almost total absence of it, and ending with a complete hallucination. The continuity is, however, not simply that of varying degrees of intensity, but of variations in the character of the process itself, so that it is by no means uncommon to find two very different forms of it concurrent in the same person. There are some who visualise well, and who also are seers of visions, who declare that the vision is not a vivid visualisation, but altogether a different phenomenon. In short, if we please to call all sensations due to external impressions "*direct*," and all others "*induced*," then there are many channels through which the "*induction*" of the latter may take place, and the channel of ordinary visualisation in the persons just mentioned is different from that through which their visions arise.

The following is a good instance of this condition. A friend writes:—

"These visions often appear with startling vividness, and so far from depending on any voluntary effort of the mind, they

[1] The details and illustrations of four other experiments with the image of a rosebud have been given me. They all vary in detail.

remain when I often wish them very much to depart, and no effort of the imagination can call them up. I lately saw a framed portrait of a face which seemed more lovely than any painting I have ever seen, and again I often see fine landscapes which bear no resemblance to any scenery I have ever looked upon. I find it difficult to define the difference between a waking vision and a mental image, although the difference is very apparent to myself. I think I can do it best in this way. If you go into a theatre and look at a scene—say of a forest by moonlight—at the back part of the stage you see every object distinctly and sufficiently illuminated (being thus unlike a mere act of memory), but it is nevertheless vague and shadowy, and you might have difficulty in telling afterwards all the objects you have seen. This resembles a mental image in point of clearness. The waking vision is like what one sees in the open street in broad daylight, when every object is distinctly impressed on the memory. The two kinds of imagery differ also as regards voluntariness, the image being entirely subservient to the will, the visions entirely independent of it. They differ also in point of suddenness, the images being formed comparatively slowly as memory recalls each detail, and fading slowly as the mental effort to retain them is relaxed, the visions appearing and vanishing in an instant. The waking visions seem quite close, filling as it were the whole head, while the mental image seems farther away in some far-off recess of the mind."

The number of sane persons who see visions no less distinctly than this correspondent is much greater than I had any idea of when I began this inquiry. I have received an interesting sketch of one, prefaced by a description of it by Mrs. Haweis. She says :—

"All my life long I have had one very constantly-recurring vision, a sight which came whenever it was dark or darkish, in bed or otherwise. It is a flight of pink roses floating in a mass from left to right, and this cloud or mass of roses is presently effaced by a flight of 'sparks' or gold speckles across them. The sparks totter or vibrate from left to right, but they fly distinctly upwards ; they are like tiny blocks, half gold, half black, rather symmetrically placed behind each other, and they are always in a hurry to efface the roses ; sometimes they have come at my call, sometimes by surprise, but they are always equally pleasing. What interests me most is that, when a child under nine, the flight of roses was light, slow, soft, close to my eyes, roses so large and brilliant and palpable that I tried to touch them ; the *scent* was overpowering, the petals perfect, with leaves peeping here and there, texture and motion all natural. They would stay a long time before the sparks came, and they occupied a large area in black space. Then the sparks came slowly flying, and generally, not always, effaced the roses at once, and every effort to retain the roses failed. Since an early age the flight of roses has annually grown smaller, swifter, and farther off, till by the time I was grown up my vision had become a speck, so instantaneous that I had hardly time to realise that it was there before the fading sparks showed that it was past. This is how they still come. The pleasure of them is past, and it always depresses me to speak of them, though I do not now, as I did when a child, connect the vision with any elevated spiritual state. But when I read Tennyson's *Holy Grail*, I wondered whether anybody else had had my vision, 'Rose-red, with beatings in it.' I may add, I was a London child who never was in the country but once, and I connect no particular flowers with that visit. I may almost say that I had never seen a rose, certainly not a quantity of them together."

A common form of vision is a phantasmagoria, or the appearance of a crowd of phantoms, sometimes hurrying past like men in a street. It is occasionally seen in broad daylight, much more often in the dark; it may be at the instant of putting out the candle, but it generally comes on when the person is in bed, preparing to sleep, but by no means yet asleep. I know no less than three men, eminent in the scientific world, who have these phantasmagoria in one form or another. It will seem curious, but it is a fact that I know of no less than five editors of very influential newspapers who experience these night visitations in a vivid form. Two of them have described the phenomena very forcibly in print, but anonymously, and two others have written on cognate experiences.

A near relative of my own saw phantasmagoria very frequently. She was eminently sane, and of such good constitution that her faculties were hardly impaired until near her death at ninety. She frequently described them to me. It gave her amusement during an idle hour to watch these faces, for their expression was always pleasing, though never strikingly beautiful. No two faces were ever alike, and no face ever resembled that of any acquaintance. When she was not well the faces usually came nearer to her, sometimes almost suffocatingly close. She never mistook them for reality, although they were very distinct. This is quite a typical case, similar in most respects to many others that I have.[1]

A notable proportion of sane persons have had not only visions, but actual hallucinations of sight, sound, or other sense, at one or more periods of their lives. I have a considerable packet of instances contributed by my personal friends, besides a large number communicated to me by other correspondents. One lady, a distinguished authoress, who was at the time a little fidgeted, but in no way over-wrought or ill, assured me that she once saw the principal character of one of her novels glide through the door straight up to her. It was about the size of a large doll, and it disappeared as suddenly as it came. Another lady, the daughter of an eminent musician, often imagines she hears her father playing. The day she told me of it the incident had again occurred. She was sitting in her room with her maid, and she asked the maid to open the door that she might hear the music better. The moment the maid got up the hallucination disappeared. Again, another lady, apparently in vigorous health, and belonging to a vigorous family, told me that during some past months she had been plagued by voices. The words were at first simple nonsense; then the word "pray" was frequently repeated; this was followed by some more or less coherent sentences of little import, and finally the voices left her. In short, the familiar hallucinations of the insane are to be met with far more frequently than is commonly supposed, among people moving in society and in good working health.

I have now nearly done with my summary of facts; it remains to make a few comments on them.

The weirdness of visions lies in their sudden appearance, in their vividness while present, and in their sudden departure. An incident in the Zoological Gardens struck me as a helpful simile. I happened to walk to the seal-pond at a moment when a sheen rested on the unbroken surface of the water. After waiting a while I became suddenly aware of the head of a seal, black, conspicuous,

[1] See some curious correspondence on this subject in the *St. James' Gazette*, Feb. 10, 15, and 20, 1882.

and motionless, just as though it had always been there, at a spot on which my eye had rested a moment previously and seen nothing. Again, after a while my eye wandered, and on its returning to the spot the seal was gone. The water had closed in silence over its head without leaving a ripple, and the sheen on the surface of the pond was as unbroken as when I first reached it. Where did the seal come from, and whither did it go? This could easily have been answered if the glare had not obstructed the view of the movements of the animal under water. As it was, a solitary link in a continuous chain of actions stood isolated from all the rest. So it is with the visions; a single stage in a series of mental processes emerges into the domain of consciousness. All that precedes and follows lies outside of it, and its character can only be inferred. We see in a general way that a condition of the presentation of visions lies in the over-sensitiveness of certain tracks or domains of brain action and the under-sensitiveness of others, certain stages in a mental process being represented very vividly in consciousness while the other stages are unfelt; also that individualism is changed to dividualism.

I do not recollect seeing it remarked that the ordinary phenomena of dreaming seem to show that partial sensitiveness is a normal condition during sleep. They do so because one of the most marked characteristics of the dreamer is the absence of common sense. He accepts wildly-incongruous visions without the slightest scepticism. Now common sense consists in the comprehension of a large number of related circumstances, and implies the simultaneous working of many parts of the brain. On the other hand, the brain is known to be imperfectly supplied with blood during sleep, and cannot therefore be at full work. It is probable enough, from hydraulic analogies, that imperfect irrigation would lead to partial irrigation, and therefore to suppression of action in some parts of the brain, and that this is really the case seems to be proved by the absence of common sense during dreams.

A convenient distinction is made between hallucinations and illusions. Hallucinations are defined as appearances wholly due to fancy; illusions, as fanciful perceptions of objects actually seen. There is also a hybrid case which depends on fanciful visions fancifully perceived. The problems we have to consider are, on the one hand, those connected with "*induced*" vision, and, on the other hand, those connected with the interpretation of vision, whether the vision be *direct* or *induced*.

It is probable that much of what passes for hallucination proper belongs in reality to the hybrid case, being an illusive interpretation of some induced visual cloud or blur. I spoke of the ever-varying patterns in the optical field ; these, under some slight functional change, may become more consciously present, and be interpreted into fantasmal appearances. Many cases could be adduced to support this view.

I will begin with illusions. What is the process by which they are established ? There is no simpler way of understanding it than by trying, as children often do, to see "faces in the fire," and to carefully watch the way in which they are first caught. Let us call to mind at the same time the experience of past illnesses, when the listless gaze wandered over the patterns on the wall-paper and the shadows of the bed-curtains, and slowly evoked the appearances of faces and figures that were not easily laid again. The process of making the faces is so rapid in health that it is difficult to analyse it without the recollection of what took place more slowly when we were weakened by illness. The first essential element in their construction is, I believe, the smallness of the area covered by the glance at any instant, so that the eye has to travel over a long track before it has visited every part of the object towards which the attention is directed generally. It is as with a plough, that must travel many miles before the whole of a small field can be tilled, but with this important difference—the plough travels methodically up and down in parallel furrows ; the eye wanders in devious curves, with abrupt bends, and the direction of its course at any instant depends on four causes : (1) on the easiest sequence of muscular motion, speaking in a general sense, (2) on idiosyncrasy, (3) on the mood, and (4) on the associations current at the moment. The effect of idiosyncrasy is excellently illustrated by the "Number-Forms," where we observe that a very special sharply-defined track of mental vision is preferred by each

individual who sees them. The influence of the mood of
the moment is shown in the curves that are felt appropriate
to the various emotions, as the lank drooping lines of grief,
which make the weeping willow so fit an emblem of it. In
constructing fire-faces it seems to me that the eye in its
wanderings tends to follow a favourite course, and it
especially dwells upon the marks that happen to coincide
with that course. It feels its way, easily diverted by
associations based on what has just been noticed, until at
last, by the unconscious practice of a system of "trial and
error," it hits upon a track that will suit—one that is easily
run over and that strings together accidental marks in a way
that happens to form a well-connected picture. This fancy
picture is then dwelt upon ; all that is incongruous with it
becomes disregarded, while all deficiencies in it are supplied
by the fantasy. The latest stages of the process might be
represented by a diorama. Three lanterns would converge
on the same screen. The first throws an image of what
the imagination will discard, the second of that which it will
retain, the third of that which it will supply. Turn on the
first and second, and the picture on the screen will be
identical with that which fell on the retina. Shut off the
first and turn on the third, and the picture will be identical
with the illusion.

Turner the painter made frequent use of a practice analogous to that of looking for fire-faces in the burning coals ;
he was known to give colours to children to daub in play on
paper, while he keenly watched for suggestive but accidental
combinations.

I have myself had frequent experience of the automatic
construction of fantastic figures, through a practice I have
somewhat encouraged for the purpose, of allowing my hand
to scribble at its own will, while I am giving my best
attention to what is being said by others, as at small com-
mittees. It is always a surprise to me to see the result
whenever I turn my thoughts on what I have been sub-
consciously doing. I can rarely recollect even a few of the
steps by which the drawings were made ; they grew piece-
meal, with some almost forgotten notice, from time to time,
of the sketch as a whole. I can trace no likeness between
what I draw and the images that present themselves to me

in dreams, and I find that a very trifling accident, such as a chance dot on the paper, may have great influence on the general character of any one of these automatic sketches.

Visions, like dreams, are often mere patchworks built up of bits of recollections. The following is one of these :—

"When passing a shop in Tottenham Court Road, I went in to order a Dutch cheese, and the proprietor (a bullet-headed man whom I had never seen before) rolled a cheese on the marble slab of his counter, asking me if that one would do. I answered 'Yes,' left the shop, and thought no more of the incident. The following evening, on closing my eyes, I saw a head detached from the body rolling about slightly on a white surface. I recognised the face, but could not remember where I had seen it, and it was only after thinking about it for some time that I identified it as that of the cheesemonger who had sold me the cheese on the previous day. I may mention that I have often seen the man since, and that I found the vision I saw was exactly like him, although if I had been asked to describe the man before I saw the vision I should have been unable to do so."

Recollections need not be combined like mosaic work ; they may be blended, on the principle of composite portraiture. I suspect that the phantasmagoria may be in some part due to blended memories ; the number of possible combinations would be practically endless, and each combination would give a new face. There would thus be no limit to the dies in the coinage of the brain.

I have found that the peculiarities of visualisation, such as the tendency to see Number-Forms, and the still rarer tendency to associate colour with sound, is strongly hereditary, and I should infer, what facts seem to confirm, that the tendency to be a seer of visions is equally so. Under these circumstances we should expect that it would be unequally developed in different races, and that a large natural gift of the visionary faculty might become characteristic not only of certain families, as among the second-sight seers of Scotland, but of certain races, as that of the Gipsies.

It happens that the mere acts of fasting, of want of sleep, and of solitary musing, are severally conducive to visions. I have myself been told of cases in which persons accidentally long deprived of food became for a brief time subject to them. One was of a pleasure party driven out to sea, and not being able to reach the coast till nightfall, at a place where they got shelter but nothing to eat. They were mentally at ease and conscious of safety, but all were troubled with visions that were half dreams and half hallucinations. The cases of visions following protracted wakefulness are well known, and I have collected a few of them myself. I have already spoken of the maddening effect of solitariness: its influence may be inferred from the recognised advantages of social amusements in the treatment of the insane. It follows that the spiritual discipline undergone for purposes of self-control and self-mortification, have also the incidental effect of producing visions. It is to be expected that these should often bear a close relation to the prevalent subjects of thought, and although they may be really no more than the products of one portion of the brain, which another portion of the same brain is engaged in contemplating, they often, through error, receive a religious sanction. This is notably the case among half-civilised races.

The number of great men who have been once, twice, or more frequently, subject to hallucinations is considerable. A list, to which it would be easy to make large additions, is given by Brierre de Boismont (*Hallucinations, etc.*, 1862), from whom I translate the following account of the star of the first Napoleon, which he heard, second-hand, from General Rapp:—

"In 1806 General Rapp, on his return from the siege of Dantzic, having occasion to speak to the Emperor, entered his study without being announced. He found him so absorbed that his entry was unperceived. The General seeing the Emperor continue motionless, thought he might be ill, and purposely made a noise. Napoleon immediately roused himself, and without any preamble, seizing Rapp by the arm, said to him, pointing to the sky, 'Look there, up there.' The General remained silent, but on being asked a second time, he answered that he perceived nothing. 'What!' replied the Emperor, 'you do not see it? It is my star, it is before you, brilliant;' then animating by degrees, he cried out, 'it has never abandoned me, I see it on all great occasions, it commands me to go forward, and it is a constant sign of good fortune to me.'"

Napoleon was no doubt a consummate actor, ready and unscrupulous in imposing on others, but I see no reason to distrust the genuineness of this particular outburst, seeing that it is not the only instance of his referring to the guidance of his star, as a literal vision and not as a mere phrase, and that his belief in destiny was notorious.

It appears that stars of this kind, so frequently spoken of in history, and so well known as a metaphor in language, are a common hallucination of the insane. Brierre de Boismont has a chapter on the stars of great men. I cannot doubt that visions of this description were in some cases the basis of that firm belief in astrology, which not a few persons of eminence formerly entertained.

The hallucinations of great men may be accounted for in part by their sharing a tendency which we have seen to be not uncommon in the human race, and which, if it happens to be natural to them, is liable to be developed in their overwrought brains by the isolation of their lives. A man in the position of the first Napoleon could have no intimate associates ; a great philosopher who explores ways of thought far ahead of his contemporaries must have an inner world in which he passes long and solitary hours. Great men may be even indebted to touches of madness for their greatness ; the ideas by which they are haunted, and to whose pursuit they devote themselves, and by which they rise to eminence, having much in common with the monomania of insanity. Striking instances of great visionaries may be mentioned, who had almost beyond doubt those very nervous seizures with which the tendency to hallucinations is intimately connected. To take a single instance, Socrates, whose *daimon* was an audible not a visual appearance, was, as has been often pointed out, subject to cataleptic seizure, standing all night through in a rigid attitude.

It is remarkable how largely the visionary temperament has manifested itself in certain periods of history and epochs of national life. My interpretation of the matter, to a certain extent, is this—That the visionary tendency is much more common among sane people than is generally suspected. In early life, it seems to be a hard lesson to an imaginative child to distinguish between the real and visionary world. If the fantasies are habitually laughed at and otherwise discouraged, the child soon acquires the power of distinguishing them ; any incongruity or nonconformity is quickly noted, the visions are found out and discredited, and are no further attended to. In this way the natural tendency to see them is blunted by repression. Therefore, when popular opinion is of a matter-of-fact kind, the seers of visions keep quiet; they do not like to be thought fanciful or mad, and they hide their experiences, which only come to light through inquiries such as these that I have been making. But let the tide of opinion change and grow favourable to supernaturalism, then the seers of visions come to the front. The faintly-perceived fantasies of ordinary persons become invested by the authority of reverend men with a claim to serious regard; they are consequently attended to and encouraged, and they increase in definition through being habitually dwelt upon. We need not suppose that a faculty previously non-existent has been suddenly evoked, but that a faculty long smothered by many in secret has been suddenly allowed freedom to express itself, and to run into extravagance owing to the removal of reasonable safeguards.

NURTURE AND NATURE.

Man is so educable an animal that it is difficult to distinguish between that part of his character which has been acquired through education and circumstance, and that which was in the original grain of his constitution. His character is exceedingly complex, even in members of the simplest and purest savage race; much more is it so in civilised races, who have long since been exempted from the full rigour of natural selection, and have become more mongrel in their breed than any other animal on the face of the earth. Different aspects of the multifarious character of man respond to different calls from without, so that the same individual, and, much more, the same race, may behave very differently at different epochs. There may have been no fundamental change of character, but a different phase or mood of it may have been evoked by special circumstances, or those persons in whom that mood is naturally dominant may through some accident have the opportunity of acting for the time as representatives of the race. The same nation may be seized by a military fervour at one period, and by a commercial one at another; they may be humbly submissive to a monarch, or become outrageous republicans. The love of art, gaiety, adventure, science, religion may be severally paramount at different times.

One of the most notable changes that can come over a nation is from a state corresponding to that of our past dark ages into one like that of the Renaissance. In the first case the minds of men are wholly taken up with routine work, and in copying what their predecessors have done; they degrade into servile imitators and submissive slaves to the past. In the second case, some circumstance or idea has finally discredited the authorities that impeded intellectual growth, and has unexpectedly revealed new possibilities. Then the mind of the nation is set free, a direction of research is given to it, and all the exploratory and hunting instincts are awakened. These sudden eras of great intellectual progress cannot be due to any alteration in the natural faculties of the race, because there has not been time for that, but to their being directed in productive channels. Most of the leisure of the men of every nation is spent in rounds of reiterated actions; if it could be spent in continuous advance along new lines of research in unexplored regions, vast progress would be sure to be made. It has been the privilege of this generation to have had fresh fields of research pointed out to them by Darwin, and to have undergone a new intellectual birth under the inspiration of his fertile genius.

A pure love of change, acting according to some law of contrast as yet imperfectly understood, especially characterises civilised man. After a long continuance of one mood he wants to throw himself into another for the pleasure of setting faculties into action that have been long disused, but not yet paralysed by disuse, and which have become fidgety for employment. He has so many opportunities for procuring change, and has so complex a nature that he easily learns to neglect a more deeply-seated feeling that innovation is wicked, and which is manifest in children and barbarians. To a civilised man the varied interests of civilisation are temptations in as many directions; changes in dress and appliances of all kinds are comparatively inexpensive to him owing to the cheapness of manufactures and their variety; change of scene is easy from the conveniences of locomotion. But a barbarian has none of these facilities: his interests are few; his dress, such as it is, is intended to stand the wear and tear of years, and all weathers; it is relatively very costly, and is an investment, one may say, of his capital rather than of his income; the invention of his people is sluggish, and their arts are few, consequently he is perforce taught to be conservative, his ideas are fixed, and he becomes scandalised even at the suggestion of change.

The difficulty of indulging in variety is incomparably greater among the rest of the animal world. If a pea-hen should take it into her head that bars would be prettier than eyes in the tail of her spouse, she could not possibly get what she wanted. It would require hundreds of generations in which the pea-hens generally concurred in the same view before sexual selection could effect the desired alteration. The feminine delight of indulging her caprice in matters of ornament is a luxury denied to the females of the brute world, and the law that rules changes in taste, if studied at all, can only be ascertained by observing the alternations of fashion in civilised communities.

There are long sequences of changes in character, which, like the tunes of a musical snuff-box, are regulated by internal mechanism. They are such as those of Shakespeare's "Seven Ages," and others due to the progress of various diseases. The lives of birds are characterised by long chains of these periodic sequences. They are mostly mute in winter, after that they begin to sing; some species are seized in the early part of the year with so strong a passion for migrating that if confined in a cage they will beat themselves to death against its bars; then follow courtship and pairing, accompanied by an access of ferocity among the males and severe fighting for the females. Next an impulse seizes them to build nests, then a desire for incubation, then one for the feeding of their young. After this a newly-arisen tendency to gregariousness groups them into large flocks, and finally they fly away to the place whence they came, goaded by a similar instinct to that which drove them forth a few months previously. These remarkable changes are mainly due to the conditions of their natures, because they persist with more or less regularity under altered circumstances. Nevertheless, they are not wholly independent of circumstance, because the period of migration, though nearly coincident in successive years, is modified to some small extent by the weather and condition of the particular year.

The interaction of nature and circumstance is very close, and it is impossible to separate them with precision. Nurture acts before birth, during every stage of embryonic and pre-embryonic existence, causing the potential faculties at the time of birth to be in some degree the effect of nurture. We need not, however, be hypercritical about distinctions; we know that the bulk of the respective provinces of nature and nurture are totally different, although the frontier between them may be uncertain, and we are perfectly justified in attempting to appraise their relative importance.

I shall begin with describing some of the principal influences that may safely be ascribed to education or other circumstances, all of which I include under the comprehensive term of Nurture.

ASSOCIATIONS.

The furniture of a man's mind chiefly consists of his recollections and the bonds that unite them. As all this is the fruit of experience, it must differ greatly in different minds according to their individual experiences. I have endeavoured to take stock of my own mental furniture in the way described in the next chapter, in which it will be seen how large a part consists of childish recollections, testifying to the permanent effect of many of the results of early education. The same fact has been strongly brought out by the replies from correspondents whom I had questioned on their mental imagery. It was frequently stated that the mental image invariably evoked by certain words was some event of childish experience or fancy. Thus one correspondent, of no mean literary and philosophical power, recollects the left hand by a mental reference to the rocking-horse which always stood by the side of the nursery wall with its head in the same direction, and had to be mounted from the side next the wall. Another, a politician, historian, and scholar, refers all his dates to the mental image of a nursery diagram of the history of the world, which has since developed huge bosses to support his later acquired information.

Our abstract ideas being mostly drawn from external experiences, their character also must depend upon the events of our individual histories. For example, the spoken words house and home must awaken ideas derived from the houses and the homes with which the hearer is, in one way or other, acquainted, and these could not be the same to persons of various social positions and places of residence. The character of our abstract ideas, therefore, depends, to a considerable degree, on our nurture.

I doubt, however, whether "abstract idea" is a correct phrase in many of the cases in which it is used, and whether "cumulative idea" would not be more appropriate. The ideal faces obtained by the method of composite portraiture appear to have a great deal in common with these so-called abstract ideas. The composite portraits consist, as was explained, of numerous superimposed pictures, forming a cumulative result in which the features that are common to all the likenesses are clearly seen; those that are common to a few are relatively faint and are more or less overlooked, while those that are peculiar to single individuals leave no sensible trace at all.

This analogy, which I pointed out in a Memoir on Generic Images,[1] has been extended and confirmed by subsequent experience of the process. One objection to my view was that our so-called generalisations are commonly no more than representative cases, our recollections being apt to be unduly influenced by particular events, and not by the totality of what we have seen ; that the reason why some one recollection has prevailed is that the case was sharply defined, or had something unusual about it, or that our frame of mind was at the time of observation susceptible to that particular kind of impression. I have had exactly the same difficulties with the composites. If one of the individual portraits has sharp outlines, or if it is

[1] "Generic Images," *Proc. Royal Institute,* Friday, April 25, 1879, partly reprinted in the Appendix.

unlike the rest, or if the illumination is temporarily strong, it will assert itself unduly in the result. The cases seem to me exactly analogous. I get over my photographic difficulty very easily by throwing the sharp portrait a little out of focus, by eliminating such portraits as have exceptional features, and by toning down the illumination to a standard intensity.

PSYCHOMETRIC EXPERIMENTS.

When we attempt to trace the first steps in each operation of our minds, we are usually baulked by the difficulty of keeping watch, without embarrassing the freedom of its action. The difficulty is much more than the common and well-known one of attending to two things at once. It is especially due to the fact that the elementary operations of the mind are exceedingly faint and evanescent, and that it requires the utmost painstaking to watch them properly. It would seem impossible to give the required attention to the processes of thought, and yet to think as freely as if the mind had been in no way preoccupied. The peculiarity of the experiments I am about to describe is that I have succeeded in evading this difficulty. My method consists in allowing the mind to play freely for a very brief period, until a couple or so of ideas have passed through it, and then, while the traces or echoes of those ideas are still lingering in the brain, to turn the attention upon them with a sudden and complete awakening; to arrest, to scrutinise them, and to record their exact appearance. Afterwards I collate the records at leisure, and discuss them, and draw conclusions. It must be understood that the second of the two ideas was never derived from the first, but always directly from the original object. This was ensured by absolutely withstanding all temptation to reverie. I do not mean that the first idea was of necessity a simple elementary thought; sometimes it was a glance down a familiar line of associations, sometimes it was a well-remembered mental attitude or mode of feeling, but I mean that it was never so far indulged in as to displace the object that had suggested it from being the primary topic of attention.

I must add, that I found the experiments to be extremely

trying and irksome, and that it required much resolution to go through with them, using the scrupulous care they demanded. Nevertheless the results well repaid the trouble. They gave me an interesting and unexpected view of the number of the operations of the mind, and of the obscure depths in which they took place, of which I had been little conscious before. The general impression they have left upon me is like that which many of us have experienced when the basement of our house happens to be under thorough sanitary repairs, and we realise for the first time the complex system of drains and gas and water pipes, flues, bell-wires, and so forth, upon which our comfort depends, but which are usually hidden out of sight, and with whose existence, so long as they acted well, we had never troubled ourselves.

The first experiments I made were imperfect, but sufficient to inspire me with keen interest in the matter, and suggested the form of procedure that I have already partly described. My first experiments were these. On several occasions, but notably on one when I felt myself unusually capable of the kind of effort required, I walked leisurely along Pall Mall, a distance of 450 yards, during which time I scrutinised with attention every successive object that caught my eyes, and I allowed my attention to rest on it until one or two thoughts had arisen through direct association with that object; then I took very brief mental note of them, and passed on to the next object. I never allowed my mind to ramble. The number of objects viewed was, I think, about 300, for I had subsequently repeated the same walk under similar conditions and endeavoured to estimate their number, with that result. It was impossible for me to recall in other than the vaguest way the numerous ideas that had passed through my mind; but of this, at least, I am sure, that samples of my whole life had passed before me, that many bygone incidents, which I never suspected to have formed part of my stock of thoughts, had been glanced at as objects too familiar to awaken the attention. I saw at once that the brain was vastly more active than I had previously believed it to be, and I was perfectly amazed at the unexpected width of the field of its everyday operations. After an interval of some days, during which I kept my mind from dwelling on my first experiences, in order that it might retain as much freshness as possible for a second experiment, I repeated the walk, and was struck just as much as before by the variety of the ideas that presented themselves, and the number of events to which they referred, about which I had never consciously occupied myself of late years.   But my admiration at the activity of the mind was seriously diminished by another observation which I then made, namely, that there had been a very great deal of repetition of thought.   The actors in my mental stage were indeed very numerous, but by no means so numerous as I had imagined.   They now seemed to be something like the actors in theatres where large processions are represented, who march off one side of the stage, and, going round by the back, come on again at the other.   I accordingly cast about for means of laying hold of these fleeting thoughts, and, submitting them to statistical analysis, to find out more about their tendency to repetition and other matters, and the method I finally adopted was the one already mentioned.   I selected a list of suitable words, and wrote them on different small sheets of paper.   Taking care to dismiss them from my thoughts when not engaged upon them, and allowing some days to elapse before I began to use them, I laid one of these sheets with all due precautions under a book, but not wholly covered by it, so that when I leaned forward I could see one of the words, being previously quite ignorant of what the word would be. Also I held a small chronograph, which I started by pressing a spring the moment the word caught my eye, and which stopped of itself the instant I released the spring; and this I did so soon as about a couple of ideas in direct association with the word had arisen in my mind.   I found that I could not manage to recollect more than two ideas with the needed precision, at least not in a general way; but sometimes several ideas occurred so nearly together that I was able to record three or even four of them, while sometimes I only managed one.   The second ideas were, as I have already said, never derived from the first, but always direct from the word itself, for I kept my attention firmly fixed on the word, and the associated ideas were seen only by a half glance.   When the two ideas had occurred,

I stopped the chronograph and wrote them down, and the time they occupied. I soon got into the way of doing all this in a very methodical and automatic manner, keeping the mind perfectly calm and neutral, but intent and, as it were, at full cock and on hair trigger, before displaying the word. There was no disturbance occasioned by thinking of the forthcoming revulsion of the mind the moment before the chronograph was stopped. My feeling before stopping it was simply that I had delayed long enough, and this in no way interfered with the free action of the mind. I found no trouble in ensuring the complete fairness of the experiment, by using a number of little precautions, hardly necessary to describe, that practice quickly suggested, but it was a most repugnant and laborious work, and it was only by strong self-control that I went through my schedule according to programme. The list of words that I finally secured was 75 in number, though I began with more. I went through them on four separate occasions, under very different circumstances, in England and abroad, and at intervals of about a month. In no case were the associations governed to any degree worth recording, by remembering what had occurred to me on previous occasions, for I found that the process itself had great influence in discharging the memory of what it had just been engaged in, and I, of course, took care between the experiments never to let my thoughts revert to the words. The results seem to me to be as trustworthy as any other statistical series that has been collected with equal care.

On throwing these results into a common statistical hotchpot, I first examined into the rate at which these associated ideas were formed. It took a total time of 660 seconds to form the 505 ideas; that is, at about the rate of 50 in a minute, or 3000 in an hour. This would be miserably slow work in reverie, or wherever the thought follows the lead of each association that successively presents itself. In the present case, much time was lost in mentally taking the word in, owing to the quiet unobtrusive way in which I found it necessary to bring it into view, so as not to distract the thoughts. Moreover, a substantive standing by itself is usually the equivalent of too abstract an idea for us to conceive properly without delay. Thus it is very difficult to get a quick conception of the word "carriage," because there are so many different kinds—two-wheeled, four-wheeled, open and closed, and all of them in so many different possible positions, that the mind possibly hesitates amidst an obscure sense of many alternatives that cannot blend together. But limit the idea to say a laudau, and the mental association declares itself more quickly. Say a laudau coming down the street to opposite the door, and an image of many blended laudaus that have done so forms itself without the least hesitation.

Next, I found that my list of 75 words gone over 4 times, had given rise to 505 ideas and 13 cases of puzzle, in which nothing sufficiently definite to note occurred within the brief maximum period of about 4 seconds, that I allowed myself to any single trial. Of these 505 only 289 were different. The precise proportions in which the 505 were distributed in quadruplets, triplets, doublets, or singles, is shown in the uppermost lines of Table I. The same facts are given under another form in the lower lines of the Table, which show how the 289 different ideas were distributed in cases of fourfold, treble, double, or single occurrences.

TABLE I.

RECURRENT ASSOCIATIONS.

Total Number of Associations.	Occurring in			
	Quadruplets.	Triplets.	Doublets.	Singles.
505	116	108	114	167
Per cent . 100	23	21	23	33

Total Number of Different Associations.	Occurring			
	Four times.	Three times.	Twice.	Once.
289	29	36	57	167
Per cent . 100	10	12	20	58

I was fully prepared to find much iteration in my ideas but had little expected that out of every hundred words twenty-three would give rise to exactly the same association in every one of the four trials; twenty-one to the same association in three out of the four, and so on, the experiments having been purposely conducted under very different conditions of time and local circumstances. This shows much less variety in the mental stock of ideas than I had expected, and makes us feel that the roadways of our minds are worn into very deep ruts. I conclude from the proved number of faint and barely conscious thoughts, and from the proved iteration of them, that the mind is perpetually travelling over familiar ways without our memory retaining any impression of its excursions. Its footsteps are so light and fleeting that it is only by such experiments as I have described that we can learn anything about them. It is apparently always engaged in mumbling over its old stores, and if any one of these is wholly neglected for a while, it is apt to be forgotten, perhaps irrecoverably. It is by no means the keenness of interest and of the attention when first observing an object, that fixes it in the recollection. We pore over the pages of a *Bradshaw*, and study the trains for some particular journey with the greatest interest; but the event passes by, and the hours and other facts which we once so eagerly considered become absolutely forgotten. So in games of whist, and in a large number of similar instances. As I understand it, the subject must have a continued living interest in order to retain an abiding place in the memory. The mind must refer to it frequently, but whether it does so consciously or unconsciously is not perhaps a matter of much importance. Otherwise, as a general rule, the recollection sinks, and appears to be utterly drowned in the waters of Lethe.

The instances, according to my personal experience, are very rare, and even those are not very satisfactory, in which some event recalls a memory that had lain *absolutely* dormant for many years. In this very series of experiments a recollection which I thought had entirely lapsed appeared under no less than three different aspects on different occasions. It was this: when I was a boy, my father, who was anxious that I should learn something of physical science, which was then never taught at school, arranged with the owner of a large chemist's shop to let me dabble at chemistry for a few days in his laboratory. ¡I had not thought of this fact, so far as I was aware, for many years ; but in scrutinising the fleeting associations called up by the various words, I traced two mental visual images (an alembic and a particular arrangement of tables and light), and one mental sense of smell (chlorine gas) to that very laboratory. I recognised that these images appeared familiar to me, but I had not thought of their origin. No doubt if some strange conjunction of circumstances had suddenly recalled those three associations at the same time, with perhaps two or three other collateral matters which may be still living in my memory, but which I no not as yet identify, a mental perception of startling vividness would be the result, and I should have falsely imagined that it had supernaturally, as it were, started into life from an entire oblivion extending over many years. Probably many persons would have registered such a case as evidence that things once perceived can never wholly vanish from the recollection, but that in the hour of death, or under some excitement, every event of a past life may reappear. To this view I entirely dissent. Forgetfulness appears absolute in the vast majority of cases, and our supposed recollections of a past life are, I believe, no more than that of a large number of episodes in it, to be reckoned perhaps in hundreds of thousands, but certainly not in tens of hundreds of thousands, that have escaped oblivion. Every one of the fleeting, half-conscious thoughts that were the subject of my experiments, admitted of being vivified by keen attention, or by some appropriate association, but I strongly suspect that ideas which have long since ceased to fleet through the brain, owing to the absence of current associations to call them up, disappear wholly. A comparison of old memories with a newly-met friend of one's boyhood, about the events we then witnessed together, show how much we had each of us forgotten. Our recollections do not tally. Actors and incidents that seem to have been of primary importance in those events to the one have been utterly forgotten by the other. The recollection of our earlier years are, in truth, very scanty, as any one will find who tries to enumerate them.

My associated ideas were for the most part due to my own unshared experiences, and the list of them would necessarily differ widely from that which another person would draw up who might repeat my experiments. Therefore one sees clearly, and I may say, one can see *measurably*, how impossible it is in a general way for two grown-up persons to lay their minds side by side together in perfect accord. The same sentence cannot produce precisely the same effect on both, and the first quick impressions that any given word in it may convey, will differ widely in the two minds.

I took pains to determine as far as feasible the dates of my life at which each of the associated ideas was first attached to the word. There were 124 cases in which identification was satisfactory, and they were distributed as in Table II.

TABLE II.

RELATIVE NUMBER OF ASSOCIATIONS FORMED AT DIFFERENT PERIODS OF LIFE.

Total number of different Associations.		Occurring								Whose first formation was in
		four times.		three times.		twice.		once.		
	per cent.		*per cent.*		*per cent.*		*per cent.*		*per cent.*	
48	*39*	12	*10*	11	*9*	9	*7*	16	*13*	boyhood and youth,
57	*46*	10	*8*	8	*7*	6	*5*	33	*26*	subsequent manhood,
19	*15*	—	—	4	*3*	1	*1*	14	*11*	quite recent events.
124	*100*	22	*18*	23	*19*	16	*13*	63	*50*	Totals.

It will be seen from the Table that out of the 48 earliest associations no less than 12, or one quarter of them, occurred in each of the four trials ; of the 57 associations first formed in manhood, 10, or about one-sixth of them, had a similar recurrence, but as to the 19 other associations first formed in quite recent times, not one of them occurred in the whole of the four trials. Hence we may see the greater fixity of the earlier associations, and might measurably determine the decrease of fixity as the date of their first formation becomes less remote.

The largeness of the number 33 in the middle entry of the last column but one, which disconcerts the run of the series, is wholly due to a visual memory of places seen in manhood. I will not speak about this now, as I shall have to refer to it farther on. Neglecting, for the moment, this unique class of occurrences, it will be seen that one-half of the associations date from the period of life before leaving college ; and it may easily be imagined that many of these refer to common events in an English education. Nay further, on looking through the list of all the associations it was easy to see how they are pervaded by purely English ideas, and especially such as are prevalent in that stratum of English society in which I was born and bred, and have subsequently lived. In illustration of this, I may mention an anecdote of a matter which greatly impressed me at the time. I was staying in a country house with a very pleasant party of young and old, including persons whose education and versatility were certainly not below the social average. One evening we played at a round game, which consisted in each of us drawing as absurd a scrawl as he or she could, representing some historical event ; the pictures were then shuffled and passed successively from hand to hand, every one writing down independently their interpretation of the picture, as to what the historical event was that the artist intended to depict by the scrawl. I was astonished at the sameness of our ideas. Cases like Canute and the waves, the Babes in the Tower, and the like, were drawn by two and even three persons at the same time, quite independently of one another, showing how narrowly we are bound by the fetters of our early education. If the figures in the above Table may be accepted as fairly correct for the world generally, it shows, still in a measurable degree, the large effect of early education in fixing our associations. It will of course be understood that I make no absurd profession of being able by these very few experiments to lay down statistical constants of universal application, but that my principal object is to show that a large class of mental phenomena, that have hitherto been too vague to lay hold of, admit of being caught by the firm grip of genuine statistical inquiry.

The results that I have thus far given are hotch-pot results. It is necessary to sort the materials somewhat before saying more about them.

After several trials I found that the associated ideas admitted of being divided into three main groups. First there is the imagined sound of words, as in verbal quotations or names of persons. This was frequently a mere parrot-like memory which acted instantaneously and in a meaningless way, just as a machine might act. In the next group there was every other kind of sense imagery; the chime of imagined bells, the shiver of remembered cold, the scent of some particular locality, and, much more frequently than all the rest put together, visual imagery. The last of the three groups contains what I will venture, for the want of a better name, to call "histrionic" representations. It includes those cases where I either act a part in imagination, or see in imagination a part acted, or, most commonly by far, where I am both spectator and all the actors at once, in an imaginary mental theatre. Thus I feel a nascent sense of some muscular action while I simultaneously witness a puppet of my brain—a part of myself—perform that action, and I assume a mental attitude appropriate to the occasion. This, in my case, is a very frequent way of generalising, indeed I rarely feel that I have secure hold of a general idea until I have translated it somehow into this form. Thus the word "abasement" presented itself to me, in one of my experiments, by my mentally placing myself in a pantomimic attitude of humiliation with half-closed eyes, bowed head, and uplifted palms, while at the same time I was aware of myself as of a mental puppet, in that position. This same word will serve to illustrate the other groups also. It so happened in connection with "abasement" that the word "David" or "King David" occurred to me on one occasion in each of three out of the four trials; also that an accidental misreading, or perhaps the merely punning association of the words "a basement," brought up on all four occasions the image of the foundations of a house that the builders had begun upon.

So much for the character of the association; next as to that of the words. I found, after the experiments were over, that the words were divisible into three distinct groups. The first contained "abbey," "aborigines," "abyss," and others that admitted of being presented under some mental image. The second group contained "abasement," "abhorrence," "ablution," etc., which admitted excellently of histrionic representation. The third group contained the more abstract words, such as "afternoon," "ability," "abnormal," which were variously and imperfectly dealt with by my mind. I give the results in the upper part of Table III., and, in order to save trouble, I have reduced them to percentages in the lower lines of the Table.

TABLE III.

COMPARISON BETWEEN THE QUALITY OF THE WORDS AND THAT OF THE IDEAS IN IMMEDIATE ASSOCIATION WITH THEM.

Number of words in each series.		Sense Imagery.	Histrionic.	Purely Names of Persons.	Verbal Phrases and Quotations.	Total.
26	"Abbey" series	46	12	32	17	107
20	"Abasement" ,,	25	26	11	17	79
29	"Afternoon" ,,	23	27	16	38	104
75						290
	"Abbey" series	43	11	30	16	100
	"Abasement" ,,	32	33	13	22	100
	"Afternoon" ,,	22	25	16	37	100

We see from this that the associations of the "abbey" series are nearly half of them in sense imagery, and these were almost always visual. The names of persons also more frequently occurred in this series than in any other. It will be recollected that in Table II. I drew attention to the exceptionally large number, 33, in the last column. It was perhaps 20 in excess of what would have been expected from the general run of the other figures. This was wholly due to visual imagery of scenes with which I was first acquainted after reaching manhood, and shows, I think, that the scenes of childhood and youth, though vividly impressed on the memory, are by no means numerous, and may be quite thrown into the background by the abundance of after experiences ; but this, as we have seen, is not the case with the other forms of association. Verbal memories of old date, such as Biblical scraps, family expressions, bits of poetry, and the like, are very numerous, and rise to the thoughts so quickly, whenever anything suggests them, that they commonly outstrip all competitors. Associations connected with the "abasement" series are strongly characterised by histrionic ideas, and by sense imagery, which to a great degree merges into a histrionic character. Thus the word "abhorrence" suggested to me, on three out of the four trials, an image of the attitude of Martha in the famous picture of the raising of Lazarus by Sebastian del Piombo in the National Gallery. She stands with averted head, doubly sheltering her face by her hands from even a sidelong view of the opened grave. Now I could not be sure how far I saw the picture as such, in my mental view, or how far I had thrown my own personality into the picture, and was acting it as actors might act a mystery play, by the puppets of my own brain, that were parts of myself. As a matter of fact, I entered it under the heading of sense imagery, but it might very properly have gone to swell the number of the histrionic entries.

The "afternoon" series suggested a great preponderance of mere catch-words, showing how slowly I was able to realise the meaning of abstractions ; the phrases intruded themselves before the thoughts became defined. It occasionally occurred that I puzzled wholly over a word, and made no entry at all ; in thirteen cases either this happened, or else after one idea had occurred the second was too confused and obscure to admit of record, and mention of it had to be omitted in the foregoing Table. These entries have forcibly shown to me the great imperfection in my generalising powers ; and I am sure that most persons would find the same if they made similar trials. Nothing is a surer sign of high intellectual capacity than the power of quickly seizing and easily manipulating ideas of a very abstract nature. Commonly we grasp them very imperfectly, and cling to their skirts with great difficulty.

In comparing the order in which the ideas presented themselves, I find that a decided precedence is assumed by the histrionic ideas, wherever they occur; that verbal associations occur first and with great quickness on many occasions, but on the whole that they are only a little more likely to occur first than second; and that imagery is decidedly more likely to be the second than the first of the associations called up by a word. In short, gesture-language appeals the most quickly to my feelings.

It would be very instructive to print the actual records at length, made by many experimenters, if the records could be clubbed together and thrown into a statistical form; but it would be too absurd to print one's own singly. They lay bare the foundations of a man's thoughts with curious distinctness, and exhibit his mental anatomy with more vividness and truth than he would probably care to publish to the world.

It remains to summarise what has been said in the foregoing memoir. I have desired to show how whole strata of mental operations that have lapsed out of ordinary consciousness, admit of being dragged into light, recorded and treated statistically, and how the obscurity that attends the initial steps of our thoughts can thus be pierced and dissipated. I then showed measurably the rate at which associations sprung up, their character, the date of their first formation, their tendency to recurrence, and their relative precedence. Also I gave an instance showing how the phenomenon of a long-forgotten scene, suddenly starting into consciousness, admitted in many cases of being explained. Perhaps the strongest of the impressions left by these experiments regards the multifariousness of the work done by the mind in a state of half-unconsciousness, and the valid reason they afford for believing in the existence of still deeper strata of mental operations, sunk wholly below the level of consciousness, which may account for such mental phenomena as cannot otherwise be explained. We gain an insight by these experiments into the marvellous number and nimbleness of our mental associations, and we also learn that they are very far indeed from being infinite in their variety. We find that our working stock of ideas is narrowly limited and that the mind continually recurs to the same instruments in conducting its operations, therefore its tracks necessarily become more defined and its flexibility diminished as age advances.

ANTECHAMBER OF CONSCIOUSNESS.

When I am engaged in trying to think anything out, the process of doing so appears to me to be this: The ideas that lie at any moment within my full consciousness seem to attract of their own accord the most appropriate out of a number of other ideas that are lying close at hand, but imperfectly within the range of my consciousness. There seems to be a presence-chamber in my mind where full consciousness holds court, and where two or three ideas are at the same time in audience, and an antechamber full of more or less allied ideas, which is situated just beyond the full ken of consciousness. Out of this antechamber the ideas most nearly allied to those in the presence-chamber appear to be summoned in a mechanically logical way, and to have their turn of audience.

The successful progress of thought appears to depend—first, on a large attendance in the antechamber; secondly, on the presence there of no ideas except such as are strictly germane to the topic under consideration; thirdly, on the justness of the logical mechanism that issues the summons. The thronging of the antechamber is, I am convinced, altogether beyond my control; if the ideas do not appear, I cannot create them, nor compel them to come. The exclusion of alien ideas is accompanied by a sense of mental effort and volition whenever the topic under consideration is unattractive, otherwise it proceeds automatically, for if an intruding idea finds nothing to cling to, it is unable to hold its place in the antechamber, and slides back again. An animal absorbed in a favourite occupation shows no sign of painful effort of attention; on the contrary, he resents interruption that solicits his attention elsewhere.

The consequence of all this is that the mind frequently

does good work without the slightest exertion. In composition it will often produce a better effect than if it acted with effort, because the essence of good composition is that the ideas should be connected by the easiest possible transitions. When a man has been thinking hard and long upon a subject, he becomes temporarily familiar with certain steps of thought, certain short cuts, and certain far-fetched associations, that do not commend themselves to the minds of other persons, nor indeed to his own at other times; therefore, it is better that his transitory familiarity with them should have come to an end before he begins to write or speak. When he returns to the work after a sufficient pause he is conscious that his ideas have settled; that is, they have lost their adventitious relations to one another, and stand in those in which they are likely to reside permanently in his own mind, and to exist in the minds of others.

Although the brain is able to do very fair work fluently in an automatic way, and though it will of its own accord strike out sudden and happy ideas, it is questionable if it is capable of working thoroughly and profoundly without past or present effort. The character of this effort seems to me chiefly to lie in bringing the contents of the antechamber more nearly within the ken of consciousness, which then takes comprehensive note of all its contents, and compels the logical faculty to test them *seriatim* before selecting the fittest for a summons to the presence-chamber.

Extreme fluency and a vivid and rapid imagination are gifts naturally and healthfully possessed by those who rise to be great orators or literary men, for they could not have become successful in those careers without it. The curious fact already alluded to of five editors of newspapers being known to me as having phantasmagoria, points to a connection between two forms of fluency, the literary and the visual. Fluency may be also a morbid faculty, being markedly increased by alcohol (as poets are never tired of telling us), and by various drugs; and it exists in delirium, insanity, and states of high emotions. The fluency of a vulgar scold is extraordinary.

In preparing to write or speak upon a subject of which the details have been mastered, I gather, after some inquiry, that the usual method among persons who have the gift of fluency is to think cursorily on topics connected with it, until what I have called the antechamber is well filled with cognate ideas. Then, to allow the ideas to link themselves in their own way, breaking the linkage continually and recommencing afresh until some line of thought has suggested itself that appears from a rapid and light glance to thread the chief topics together. After this the connections are brought step by step fully into consciousness, they are short-circuited here and extended there, as found advisable until a firm connection is found to be established between all parts of the subject. After this is done the mental effort is over, and the composition may proceed fluently in an automatic way. Though this, I believe, is a usual way, it is by no means universal, for there are very great differences in the conditions under which different persons compose most readily. They seem to afford as good evidence of the variety of mental and bodily constitutions as can be met with in any other line of inquiry.

It is very reasonable to think that part at least of the inward response to spiritual yearnings is of similar origin to the visions, thoughts, and phrases that arise automatically when the mind has prepared itself to receive them. The devout man attunes his mind to holy ideas, he excludes alien thoughts, and he waits and watches in stillness. Gradually the darkness is lifted, the silence of the mind is broken, and the spiritual responses are heard in the way so often described by devout men of all religions. This seems to me precisely analogous to the automatic presentation of ordinary ideas to orators and literary men, and to the visions of which I spoke in the chapter on that subject. Dividuality replaces individuality, and one portion of the mind communicates with another portion as with a different person.

Some persons and races are naturally more imaginative than others, and show their visionary tendency in every one of the respects named. They are fanciful, oratorical, poetical, and credulous. The "enthusiastic" faculties all seem to hang together; I shall recur to this in the chapter on enthusiasm.

I have already pointed out the existence of a morbid

form of piety : there is also a morbid condition of apparent inspiration to which imaginative women are subject, especially those who suffer more or less from hysteria. It is accompanied in a very curious way, familiar to medical men, by almost incredible acts of deceit. It is found even in ladies of position apparently above the suspicion of vulgar fraud, and seems associated with a strange secret desire to attract notice. Ecstatics, seers of visions, and devout fasting girls who eat on the sly, often belong to this category.

EARLY SENTIMENTS.

The child is passionately attached to his home, then to his school, his country, and religion ; yet how entirely the particular home, school, country, and religion are a matter of accident! He is born prepared to attach himself as a climbing plant is naturally disposed to climb, the kind of stick being of little importance. The models upon whom the child or boy forms himself are the boys or men whom he has been thrown amongst, and whom from some incidental cause he may have learned to love and respect. The every-day utterances, the likes and dislikes of his parents, their social and caste feelings, their religious persuasions are absorbed by him ; their views or those of his teachers become assimilated and made his own. If a mixed marriage should have taken place, and the father should die while the children are yet young, and if a question arise between the executors of his will and the mother as to the religious education of the children, application is made as a matter of course to the Court of Chancery, who decide that the children shall be brought up as Protestants or as Catholics as the case may be, or the sons one way and the daughters the other ; and they are, and usually remain so afterwards when free to act for themselves.

It is worthy of note that many of the deaf-mutes who are first taught to communicate freely with others after they had passed the period of boyhood, and are asked about their religious feelings up to that time, are reported to tell the same story. They say that the meaning of the church service whither they had accompanied their parents, and of the kneeling to pray, had been absolutely unintelligible, and a standing puzzle to them. The ritual touched no chord in their untaught natures that responded in unison. Very much of what we fondly look upon as a natural religious sentiment is purely traditional.

The word religion may fairly be applied to any group of sentiments or persuasions that are strong enough to bind us to do that which we intellectually may acknowledge to be our duty, and the possession of some form of religion in this larger sense of the word is of the utmost importance to moral stability. The sentiments must be strong enough to make us ashamed at the mere thought of committing, and distressed during the act of committing any untruth, or any uncharitable act, or of neglecting what we feel to be right, in order to indulge in laziness or gratify some passing desire. So long as experience shows the religion to be competent to produce this effect, it seems reasonable to believe that the particular dogma is comparatively of little importance. But as the dogma or sentiments, whatever they be, if they are not naturally instinctive, must be ingrained in the character to produce their full effect, they should be instilled early in life and allowed to grow unshaken until their roots are firmly fixed. The consciousness of this fact makes the form of religious teach'ng in every church and creed identical in one important particular though its substance may vary in every respect. In subjects unconnected with sentiment, the freest inquiry and the fullest deliberation are required before it is thought decorous to form a final opinion; but wherever sentiment is involved, and especially in questions of religious dogma, about which there is more sentiment and more difference of opinion among wise, virtuous, and truth-seeking men than about any other subject whatever, free inquiry is peremptorily discouraged. The religious instructor in every creed is one who makes it his profession to saturate his pupils with prejudice. A vast and perpetual clamour arises from the pulpits of endless proselytising sects throughout this great empire, the priests of all of them crying with one consent, "This is the way, shut your ears to the words of those who teach differently; don't look at their books, do not even mention their names except to scoff at them;

they are damnable. Have faith in what I tell you, and save your souls!" In which of these conflicting doctrines are we to place our faith if we are not to hear all sides, and to rely upon our own judgment in the end? Are we to understand that it is the duty of man to be credulous in accepting whatever the priest in whose neighbourhood he happens to reside may say? Is it to believe whatever his parents may have lovingly taught him? There are a vast number of foolish men and women in the world who marry and have children, and because they deal lovingly with their children it does not at all follow that they can instruct them wisely. Or is it to have faith in what the wisest men of all ages have found peace in believing? The Catholic phrase, "*quod semper quod ubique quod omnibus*"—"that which has been believed at all times, in all places, and by all men"—has indeed a fine rolling sound, but where is the dogma that satisfies its requirements? Or is it, such and such really good and wise men with whom you are acquainted, and whom, it may be, you have the privilege of knowing, have lived consistent lives through the guidance of these dogmas, how can you who are many grades their inferior in good works, in capacity and in experience, presume to set up your opinion against theirs? The reply is, that it is a matter of history and notoriety that other very good, capable, and inexperienced men have led and are leading consistent lives under the guidance of totally different dogmas, and that some of them a few generations back would have probably burned your modern hero as a heretic if he had lived in their times and they could have got hold of him. Also, that men, however eminent in goodness, intellect, and experience, may be deeply prejudiced, and that their judgment in matters where their prejudices are involved cannot thenceforward be trusted. Watches, as electricians know to their cost, are liable to have their steel work accidentally magnetised, and the best chronometer under those conditions can never again be trusted to keep correct time.

Lastly, we are told to have faith in our conscience? well we know now a great deal more about conscience than formerly. Ethnologists have studied the manifestations of conscience in different people, and do not find that they are consistent. Conscience is now known to be partly transmitted by inheritance in the way and under the conditions clearly explained by Mr. Darwin, and partly to be an unsuspected result of early education. The value of inherited conscience lies in its being the organised result of the social experiences of many generations, but it fails in so far as it expresses the experience of generations whose habits differed from our own. The doctrine of evolution shows that no race can be in perfect harmony with its surroundings; the latter are continually changing, while the organism of the race hobbles after, vainly trying to overtake them. Therefore the inherited part of conscience cannot be an infallible guide, and the acquired part of it may, under the influence of dogma, be a very bad one. The history of fanaticism shows too clearly that this is not only a theory but a fact. Happy the child, especially in these inquiring days, who has been taught a religion that mainly rests on the moral obligations between man and man in domestic and national life, and which, so far as it is necessarily dogmatic, rests chiefly upon the proper interpretation of facts about which there is no dispute,—namely, on those habitual occurrences which are always open to observation, and which form the basis of so-called natural religion.

It would be instructive to make a study of the working religion of good and able men of all nations, in order to discover the real motives by which they were severally animated,—men, I mean, who had been tried by both prosperity and adversity, and had borne the test; who, while they led lives full of interest to themselves, were beloved by their own family, noted among those with whom they had business relations for their probity and conciliatory ways, and honoured by a wider circle for their unselfish furtherance of the public good. Such men exist of many faiths and in many races.

Another interesting and cognate inquiry would be into the motives that have sufficed to induce men who were leading happy lives, to meet death willingly at a time when they were not particularly excited. Probably the number of instances to be found, say among Mussulmans, who are firm believers in the joys of Mahomet's Paradise, would not be more numerous than among the Zulus, who have no belief in any paradise at all, but are influenced by martial honour and patriotism. There is an Oriental phrase, as I have been told, that the fear of the inevitable approach of death is a European malady.

Terror at any object is quickly taught if it is taught consistently, whether the terror be reasonable or not. There are few more stupid creatures than fish, but they notoriously soon learn to be frightened at any newly-introduced method of capture, say by an artificial fly, which, at first their comrades took greedily. Some one fish may have seen others caught, and have learned to take fright at the fly. Whenever he saw it again he would betray his terror by some instinctive gesture, which would be seen and understood by others, and so instruction in distrusting the fly appears to spread.

All gregarious animals are extremely quick at learning terrors from one another. It is a condition of their existence that they should do so, as was explained at length in a previous chapter. Their safety lies in mutual intelligence and support. When most of them are browsing a few are always watching, and at the least signal of alarm the whole herd takes fright simultaneously. Gregarious animals are quickly alive to their mutual signals; it is beautiful to watch great flocks of birds as they wheel in their flight and suddenly show the flash of all their wings against the sky, as they simultaneously and suddenly change their direction.

Much of the tameness or wildness of an animal's character is probably due to the placidity or to the frequent starts of alarm of the mother while she was rearing it. I was greatly struck with some evidence I happened to meet with, of the pervading atmosphere of alarm and suspicion in which the children of criminal parents are brought up, and which, in combination with their inherited disposition, makes them, in the opinion of many observers, so different to other children. The evidence of which I speak lay in the tone of letters sent by criminal parents to their children, who were inmates of the Princess Mary Village Homes, from which I had the opportunity, thanks to the kindness of the Superintendent, Mrs. Meredith, of hearing and seeing extracts. They were full of such phrases as " Mind you do not say anything about this," though the matters referred to were, to all appearance, unimportant.

The writings of Dante on the horrible torments of the damned, and the realistic pictures of the same subject in frescoes and other pictures of the same date, showing the flames and the flesh hooks and the harrows, indicate the transforming effect of those cruel times, fifteen generations ago, upon the disposition of men. Revenge and torture had been so commonly practised by rulers that they seemed to be appropriate attributes of every high authority, and the artists of those days saw no incongruity in supposing that a supremely powerful master, however beneficent he might be, would make the freest use of them.

Aversion is taught as easily as terror, when the object of it is neutral and not especially attractive to an unprejudiced taste. I can testify in my own person to the somewhat rapidly-acquired and long-retained fancies concerning the clean and unclean, upon which Jews and Mussulmans lay such curious stress. It was the result of my happening to spend a year in the East, at an age when the brain is very receptive of new ideas, and when I happened to be much impressed by the nobler aspects of Mussulman civilisation, especially, I may say, with the manly conformity of their every-day practice to their creed, which contrasts sharply with what we see among most Europeans, who profess extreme unworldliness and humiliation on one day of the week, and act in a worldly and masterful manner during the remaining six. Although many years have passed since that time, I still find the old feelings in existence—for instance, that of looking on the left hand as unclean.

It is difficult to an untravelled Englishman, who has not had an opportunity of throwing himself into the spirit of the East, to credit the disgust and detestation that numerous every-day acts, which appear perfectly harmless to his countrymen, excite in many Orientals.

To conclude, the power of nurture is very great in implanting sentiments of a religious nature, of terror and of aversion, and in giving a fallacious sense of their being natural instincts. But it will be observed that the circumstances from which these influences proceed, affect large classes simultaneously, forming a kind of atmosphere in which every member of them passes his life. They produce the cast of mind that distinguishes an Englishman from a foreigner, and one class of Englishman from another, but they have little influence in creating the differences that exist between individuals of the same class.

HISTORY OF TWINS.

The exceedingly close resemblance attributed to twins has been the subject of many novels and plays, and most persons have felt a desire to know upon what basis of truth those works of fiction may rest. But twins have a special claim upon our attention; it is, that their history affords means of distinguishing between the effects of tendencies received at birth, and of those that were imposed by the special circumstances of their after lives. The objection to statistical evidence in proof of the inheritance of peculiar faculties has always been: "The persons whom you compare may have lived under similar social conditions and have had similar advantages of education, but such prominent conditions are only a small part of those that determine the future of each man's life. It is to trifling accidental circumstances that the bent of his disposition and his success are mainly due, and these you leave wholly out of account—in fact, they do not admit of being tabulated, and therefore your statistics, however plausible at first sight, are really of very little use." No method of inquiry which I had previously been able to carry out—and I have tried many methods—is wholly free from this objection. I have therefore attacked the problem from the opposite side, seeking for some new method by which it would be possible to weigh in just scales the effects of Nature and Nurture, and to ascertain their respective shares in framing the disposition and intellectual ability of men. The life-history of twins supplies what I wanted. We may begin by inquiring about twins who were closely alike in boyhood and youth, and who were educated together for many years, and learn whether they subsequently grew unlike, and, if so, what the main causes were which, in the opinion of the family, produced the dissimilarity. In this way we can obtain direct evidence of the kind we want. Again, we may obtain yet more valuable evidence by a converse method. We can inquire into the history of twins who were exceedingly unlike in childhood, and learn how far their characters became assimilated under the influence of identical nurture, isasmuch as they had the same home, the same teachers, the same associates, and in every other respect the same surroundings.

My materials were obtained by sending circulars of inquiry to persons who were either twins themselves or near relations of twins. The printed questions were in thirteen groups; the last of them asked for the addresses of other twins known to the recipient, who might be likely to respond if I wrote to them. This happily led to a continually widening circle of correspondence, which I pursued until enough material was accumulated for a general reconnaisance of the subject.

There is a large literature relating to twins in their purely surgical and physiological aspect. The reader interested in this should consult *Die Lehre von den Zwillingen*, von L. Kleinwächter, Prag. 1871. It is full of references, but it is also unhappily disfigured by a number of numerical misprints, especially in page 26. I have not found any book that treats of twins from my present point of view.

The reader will easily understand that the word "twins" is a vague expression, which covers two very dissimilar events—the one corresponding to the progeny of animals that usually bear more than one at a birth, each of the progeny being derived from a separate ovum, while the other event is due to the development of two germinal spots in the same ovum. In the latter case they are enveloped in the same membrane, and all such twins are found invariably to be of the same sex. The consequence of this is, that I find a curious discontinuity in my results. One would have expected that twins would commonly be found to possess a certain average likeness to one another; that a few would greatly exceed that average likeness, and a few would greatly fall short of it. But this is not at all the case. Extreme similarity and extreme dissimilarity between twins of the same sex are nearly as common as moderate resemblance. When the twins are a boy and a girl, they are never closely alike ; in fact, their origin is never due to the development of two germinal spots in the same ovum.

I received about eighty returns of cases of close similarity, thirty-five of which entered into many instructive details. In a few of these not a single point of difference could be specified. In the remainder, the colour of the hair and eyes were almost always identical ; the height, weight, and strength were nearly so. Nevertheless, I have a few cases of a notable difference in height, weight, and strength, although the resemblance was otherwise very near. The manner and personal address of the thirty-five pairs of twins are usually described as very similar, but accompanied by a slight difference of expression, familiar to near relatives, though unperceived by strangers. The intonation of the voice when speaking is commonly the same, but it frequently happens that the twins sing in different keys. Most singularly, the one point in which similarity is rare is the handwriting. I cannot account for this, considering how strongly handwriting runs in families, but I am sure of the fact. I have only one case in which nobody, not even the twins themselves, could distinguish their own notes of lectures, etc. ; barely two or three in which the handwriting was undistinguishable by others, and only a few in which it was described as closely alike. On the other hand, I have many in which it is stated to be unlike, and some in which it is alluded to as the only point of difference. It would appear that the handwriting is a very delicate test of difference in organisation—a conclusion which I commend to the notice of enthusiasts in the art of discovering character by the handwriting.

One of my inquiries was for anecdotes regarding mistakes made between the twins by their near relatives. The replies are numerous, but not very varied in character. When the twins are children, they are usually distinguished by ribbons tied round the wrist or neck ; nevertheless the one is sometimes fed, physicked, and whipped by mistake for the other, and the description of these little domestic catastrophes was usually given by the mother, in a phraseology that is somewhat touching by reason of its seriousness. I have one case in which a doubt remains whether the children were not changed in their bath, and the presumed A is not really B, and *vice versâ*. In another case, an artist was engaged on the portraits of twins who were between three and four years of age ; he had to lay aside his work for three weeks, and, on resuming it, could not tell to which child the respective likenesses he had in hand belonged. The mistakes become less numerous on the part of the mother during the boyhood and girlhood of the twins, but are almost as frequent as before on the part of strangers. I have many instances of tutors being unable to distinguish their twin pupils. Two girls used regularly to impose on their music teacher when one of them wanted a whole holiday ; they had their lessons at separate hours, and the one girl sacrificed herself to receive two lessons on the same day, while the other one enjoyed herself from morning to evening. Here is a brief and comprehensive account :—

"Exactly alike in all, their schoolmasters never could tell them apart; at dancing parties they constantly changed partners without discovery; their close resemblance is scarcely diminished by age."

The following is a typical schoolboy anecdote :—

"Two twins were fond of playing tricks, and complaints were frequently made ; but the boys would never own which was the guilty one, and the complainants were never certain which of the two he was. One head master used to say he would never flog the innocent for the guilty, and another used to flog both."

No less than nine anecdotes have reached me of a twin seeing his or her reflection in a looking-glass, and addressing it in the belief it was the other twin in person.

I have many anecdotes of mistakes when the twins were nearly grown up. Thus :—

"Amusing scenes occurred at college when one twin came to visit the other ; the porter on one occasion refusing to let the visitor out of the college gates, for, though they stood side by side, he professed ignorance as to which he ought to allow to depart."

Children are usually quick in distinguishing between their

parent and his or her twin; but I have two cases to the contrary. Thus, the daughter of a twin says:—

"Such was the marvellous similarity of their features, voice, manner, etc., that I remember, as a child, being very much puzzled, and I think, had my aunt lived much with us, I should have ended by thinking I had two mothers."

In the other case, a father who was a twin, remarks of himself and his brother:—

"We were extremely alike, and are so at this moment, so much so that our children up to five and six years old did not know us apart."

I have four or five instances of doubt during an engagement of marriage. Thus:—

"A married first, but both twins met the lady together for the first time, and fell in love with her there and then. A managed to see her home and to gain her affection, though B went sometimes courting in his place, and neither the lady nor her parents could tell which was which."

I have also a German letter, written in quaint terms, about twin brothers who married sisters, but could not easily be distinguished by them.[1] In the well-known novel by Mr. Wilkie Collins of *Poor Miss Finch*, the blind girl distinguishes the twin she loves by the touch of his hand, which gives her a thrill that the touch of the other brother does not. Philosophers have not, I believe, as yet investigated the conditions of such thrills; but I have a case in which Miss Finch's test would have failed. Two persons, both friends of a certain twin lady, told me that she had frequently remarked to them that "kissing her twin sister was not like kissing her other sisters, but like kissing herself—her own hand, for example."

It would be an interesting experiment for twins who were closely alike to try how far dogs could distinguish them by scent.

[1] I take this opportunity of withdrawing an anecdote, happily of no great importance, published in *Men of Science*, p. 14, about a man personating his twin brother for a joke at supper, and not being discovered by his wife. It was told me on good authority; but I have reason to doubt the fact, as the story is not known to the son of one of the twins. However, the twins in question were extraordinarily alike, and I have many anecdotes about them sent me by the latter gentleman.

I have a few anecdotes of strange mistakes made between twins in adult life. Thus, an officer writes :—

" On one occasion when I returned from foreign service my father turned to me and said, ' I thought you were in London,' thinking I was my brother—yet he had not seen me for nearly four years—our resemblance was so great."

The next and last anecdote I shall give is, perhaps, the most remarkable of those I have ; it was sent me by the brother of the twins, who were in middle life at the time of its occurrence :—

" A was again coming home from India, on leave ; the ship did not arrive for some days after it was due ; the twin brother B had come up from his quarters to receive A, and their old mother was very nervous. One morning A rushed in saying, ' Oh, mother, how are you ? ' Her answer was, ' No, B, it's a bad joke ; you know how anxious I am ! ' and it was a little time before A could persuade her that he was the real man."

Enough has been said to prove that an extremely close personal resemblance frequently exists between twins of the same sex; and that, although the resemblance usually diminishes as they grow into manhood and womanhood, some cases occur in which the diminution of resemblance is hardly perceptible. It must be borne in mind that it is not necessary to ascribe the divergence of development, when it occurs, to the effect of different nurtures, but it is quite possible that it may be due to the late appearance of qualities inherited at birth, though dormant in early life, like gout. To this I shall recur.

There is a curious feature in the character of the resemblance between twins, which has been alluded to by a few correspondents ; it is well illustrated by the following quotations. A mother of twins says :—

" There seemed to be a sort of interchangeable likeness in expression, that often gave to each the effect of being more like his brother than himself."

Again, two twin brothers, writing to me, after analysing their points of resemblance, which are close and numerous, and pointing out certain shades of difference, add—

" These seem to have marked us through life, though for a while, when we were first separated, the one to go to business, and the other to college, our respective characters were inverted; we both think that at that time we each ran into the character of the other. The proof of this consists in our own recollections, in our correspondence by letter, and in the views which we then took of matters in which we were interested."

In explanation of this apparent interchangeableness, we must recollect that no character is simple, and that in twins who strongly resemble each other, every expression in the one may be matched by a corresponding expression in the other, but it does not follow that the same expression should be the prevalent one in both cases. Now it is by their prevalent expressions that we should distinguish between the twins; consequently when one twin has temporarily the expression which is the prevalent one in his brother, he is apt to be mistaken for him. There are also cases where the development of the two twins is not strictly *pari passu;* they reach the same goal at the same time, but not by identical stages. Thus: A is born the larger, then B overtakes and surpasses A, and is in his turn overtaken by A, the end being that the twins, on reaching adult life, are of the same size. This process would aid in giving an interchangeable likeness at certain periods of their growth, and is undoubtedly due to nature more frequently than to nurture.

Among my thirty-five detailed cases of close similarity, there are no less than seven in which both twins suffered from some special ailment or had some exceptional peculiarity. One twin writes that she and her sister "have both the defect of not being able to come downstairs quickly, which, however, was not born with them, but came on at the age of twenty." Three pairs of twins have peculiarities in their fingers; in one case it consists in a slight congenital flexure of one of the joints of the little finger; it was inherited from a grandmother, but neither parents, nor brothers, nor sisters show the least trace of it. In another case the twins have a peculiar way of bending the fingers, and there was a faint tendency to the same peculiarity in the mother, but in her alone of all the family. In a third case, about which I made a few inquiries, which is given by Mr. Darwin, but is not included in my returns, there was no known family tendency to the peculiarity which was observed in the twins of having a crooked little finger. In another pair of twins, one was born ruptured, and the other became so at six months old. Two twins at the age of twenty-three were attacked by toothache, and the same tooth had to be extracted in each case. There are curious and close correspondences mentioned in the falling off of the hair. Two cases are mentioned of death from the same disease; one of which is very affecting. The outline of the story was that the twins were closely alike and singularly attached, and had identical tastes; they both obtained Government clerkships, and kept house together, when one sickened and died of Bright's disease, and the other also sickened of the same disease and died seven months later.

Both twins were apt to sicken at the same time in no less than nine out of the thirty-five cases. Either their illnesses, to which I refer, were non-contagious, or, if contagious, the twins caught them simultaneously; they did not catch them the one from the other. This implies so intimate a constitutional resemblance, that it is proper to give some quotations in evidence. Thus, the father of two twins says:—

"Their general health is closely alike; whenever one of them has an illness, the other invariably has the same within a day or two, and they usually recover in the same order. Such has been the case with whooping-cough, chicken-pox, and measles; also with slight bilious attacks, which they have successively. Latterly, they had a feverish attack at the same time."

Another parent of twins says:—

"If anything ails one of them, identical symptoms *nearly always* appear in the other; this has been singularly visible in two instances during the last two months. Thus, when in London, one fell ill with a violent attack of dysentery, and within twenty-four hours the other had precisely the same symptoms."

A medical man writes of twins with whom he is well acquainted:—

"Whilst I knew them, for a period of two years, there was not the slightest tendency towards a difference in body or mind; external influences seemed powerless to produce any dissimilarity."

The mother of two other twins, after describing how they were ill simultaneously up to the age of fifteen, adds, that they shed their first milk-teeth within a few hours of each other.

Trousseau has a very remarkable case (in the chapter on Asthma) in his important work *Clinique Médicale.* (In the edition of 1873 it is in vol. ii. p. 473.) It was quoted at length in the original French, in Mr. Darwin's *Variation under Domestication,* vol. ii. p. 252. The following is a translation:—

"I attended twin brothers so extraordinarily alike, that it was impossible for me to tell which was which, without seeing them side by side. But their physical likeness extended still deeper, for they had, so to speak, a yet more remarkable pathological resemblance. Thus, one of them, whom I saw at the Néothermes at Paris, suffering from rheumatic ophthalmia, said to me, 'At this instant my brother must be having an ophthalmia like mine;' and, as I had exclaimed against such an assertion, he showed me a few days afterwards a letter just received by him from his brother, who was at that time at Vienna, and who expressed himself in these words—'I have my ophthalmia; you must be having yours.' However singular this story may appear, the fact is none the less exact; it has not been told to me by others, but I have seen it myself; and I have seen other analogous cases in my practice. These twins were also asthmatic, and asthmatic to a frightful degree. Though born in Marseilles, they were never able to stay in that town, where their business affairs required them to go, without having an attack. Still more strange, it was sufficient for them to get away only as far as Toulon in order to be cured of the attack caught at Marseilles. They travelled continually, and in all countries, on business affairs, and they remarked that certain localities were extremely hurtful to them, and that in others they were free from all asthmatic symptoms."

I do not like to pass over here a most dramatic tale in the *Psychologie Morbide* of Dr. J. Moreau (de Tours), Médecin de l'Hospice de Bicêtre. Paris, 1859, p. 172. He speaks " of two twin brothers who had been confined, on account of monomania, at Bicêtre ":—

"Physically the two young men are so nearly alike that the one is easily mistaken for the other. Morally, their resemblance is no less complete, and is most remarkable in its details. Thus, their dominant ideas are absolutely the same. They both consider themselves subject to imaginary persecutions ; the same enemies have sworn their destruction, and employ the same means to effect it. Both have hallucinations of hearing. They are both of them melancholy and morose ; they never address a word to anybody, and will hardly answer the questions that others address to them. They always keep apart, and never communicate with one another. An extremely curious fact which has been frequently noted by the superintendents of their section of the hospital, and by myself, is this : From time to time, at very irregular intervals of two, three, and many months, without appreciable cause, and by the purely spontaneous effect of their illness, a very marked change takes place in the condition of the two brothers. Both of them, at the same time, and often on the same day, rouse themselves from their habitual stupor and prostration ; they make the same complaints, and they come of their own accord to the physician, with an urgent request to be liberated. I have seen this strange thing occur, even when they were some miles apart, the one being at Bicêtre, and the other living at Saint-Anne."

I sent a copy of this passage to the principal authorities among the physicians to the insane in England, asking if they had ever witnessed any similar case. In reply, I have received three noteworthy instances, but none to be compared in their exact parallelism with that just given. The details of these three cases are painful, and it is not necessary to my general purpose that I should further allude to them.

There is another curious French case of insanity in twins, which was pointed out to me by Sir James Paget, described by Dr. Baume in the *Annales Médico-Psychologiques*, 4 série, vol. i., 1863, p. 312, of which the following is an abstract. The original contains a few more details, but is too long to quote : François and Martin, fifty years of age, worked as railroad contractors between Quimper and Châteaulin. Martin had twice slight attacks of insanity. On January 15 a box was robbed in which the twins had deposited their savings. On the night of January 23–24 both François (who lodged at Quimper) and Martin (who lived with his wife and children at St. Lorette, two leagues from Quimper) had the same dream at the same hour, three a.m., and both awoke with a violent start, calling out, "I have caught the thief ! I have caught the thief ! they are doing mischief to my brother !" They were both of them extremely agitated, and gave way to similar extravagances, dancing and leaping.

Martin sprang on his grandchild, declaring that he was the thief, and would have strangled him if he had not been prevented; he then became steadily worse, complained of violent pains in his head, went out of doors on some excuse, and tried to drown himself in the river Steir, but was forcibly stopped by his son, who had watched and followed him. He was then taken to an asylum by gendarmes, where he died in three hours. François, on his part, calmed down on the morning of the 24th, and employed the day in inquiring about the robbery. By a strange chance, he crossed his brother's path at the moment when the latter was struggling with the gendarmes; then he himself became maddened, giving way to extravagant gestures and using incoherent language (similar to that of his brother). He then asked to be bled, which was done, and afterwards, declaring himself to be better, went out on the pretext of executing some commission, but really to drown himself in the River Steir, which he actually did, at the very spot where Martin had attempted to do the same thing a few hours previously.

The next point which I shall mention in illustration of the extremely close resemblance between certain twins is the similarity in the association of their ideas. No less than eleven out of the thirty-five cases testify to this. They make the same remarks on the same occasion, begin singing the same song at the same moment, and so on; or one would commence a sentence, and the other would finish it. An observant friend graphically described to me the effect produced on her by two such twins whom she had met casually. She said: "Their teeth grew alike, they spoke alike and together, and said the same things, and seemed just like one person." One of the most curious anecdotes that I have received concerning this similarity of ideas was that one twin, A, who happened to be at a town in Scotland, bought a set of champagne glasses which caught his attention, as a surprise for his brother B; while, at the same time, B, being in England, bought a similar set of precisely the same pattern as a surprise for A. Other anecdotes of a like kind have reached me about these twins.

The last point to which I shall allude regards the tastes and dispositions of the thirty-five pairs of twins. In sixteen cases—that is, in nearly one-half of them—these were described as closely similar; in the remaining nineteen they were much alike, but subject to certain named differences. These differences belonged almost wholly to such groups of qualities as these: The one was the more vigorous, fearless, energetic; the other was gentle, clinging, and timid; or the one was more ardent, the other more calm and placid; or again, the one was the more independent, original, and self-contained; the other the more generous, hasty, and vivacious. In short, the difference was that of intensity or energy in one or other of its protean forms; it did not extend more deeply into the structure of the characters. The more vivacious might be subdued by ill health, until he assumed the character of the other; or the latter might be raised by excellent health to that of the former. The difference was in the key-note, not in the melody.

It follows from what has been said concerning the similar dispositions of the twins, the similarity in the associations of their ideas, of their special ailments, and of their illnesses generally, that the resemblances are not superficial, but extremely intimate. I have only two cases of a strong bodily resemblance being accompanied by mental diversity, and one case only of the converse kind. It must be remembered that the conditions which govern extreme likeness between twins are not the same as those between ordinary brothers and sisters, and that it would be incorrect to conclude from what has just been said about the twins that mental and bodily likeness are invariably co-ordinate, such being by no means the case.

We are now in a position to understand that the phrase "close similarity" is no exaggeration, and to realise the value of the evidence I am about to adduce. Here are thirty-five cases of twins who were "closely alike" in body and mind when they were young, and who have been reared exactly alike up to their early manhood and womanhood. Since then the conditions of their lives have changed; what change of Nurture has produced the most variation?

It was with no little interest that I searched the records of the thirty-five cases for an answer; and they gave an answer that was not altogether direct, but it was distinct, and not at all what I had expected. They showed me that in some cases the resemblance of body and mind had continued unaltered up to old age, notwithstanding very different conditions of life; and they showed in the other cases that the parents ascribed such dissimilarity as there was, wholly or almost wholly to some form of illness. In four cases it was scarlet fever; in a fifth, typhus; in a sixth, a slight effect was ascribed to a nervous fever; in a seventh it was the effect of an Indian climate; in an eighth, an illness (unnamed) of nine months' duration; in a ninth, varicose veins; in a tenth, a bad fracture of the leg, which prevented all active exercise afterwards; and there were three additional instances of undefined forms of ill health. It will be sufficient to quote one of the returns; in this the father writes:

"At birth they were *exactly* alike, except that one was born with a bad varicose affection, the effect of which had been to prevent any violent exercise, such as dancing or running, and, as she has grown older, to make her more serious and thoughtful. Had it not been for this infirmity, I think the two would have been as exactly alike as it is possible for two women to be, both mentally and physically; even now they are constantly mistaken for one another."

In only a very few cases is some allusion made to the dissimilarity being partly due to the combined action of many small influences, and in none of the thirty-five cases is it largely, much less wholly, ascribed to that cause. In not a single instance have I met with a word about the growing dissimilarity being due to the action of the firm free-will of one or both of the twins, which had triumphed over natural tendencies; and yet a large proportion of my correspondents happen to be clergymen, whose bent of mind is opposed, as I feel assured from the tone of their letters, to a necessitarian view of life.

It has been remarked that a growing diversity between twins may be ascribed to the tardy development of naturally diverse qualities; but we have a right, upon the evidence I have received, to go farther than this. We have seen that a few twins retain their close resemblance through life; in other words, instances do exist of an apparently thorough similarity of nature, in which such difference of external circumstances as may be consistent with the ordinary conditions of the same social rank and country do not create dissimilarity. Positive evidence, such as this, cannot be outweighed by any amount of negative evidence. Therefore, in those cases where there is a growing diversity, and where no external cause can be assigned either by the twins themselves or by their family for it, we may feel sure that it must be chiefly or altogether due to a want of thorough similarity in their nature. Nay, further, in some cases it is distinctly affirmed that the growing dissimilarity can be accounted for in no other way. We may, therefore, broadly conclude that the only circumstance, within the range of those by which persons of similar conditions of life are affected, that is capable of producing a marked effect on the character of adults, is illness or some accident which causes physical infirmity. The twins who closely resembled each other in childhood and early youth, and were reared under not very dissimilar conditions, either grow unlike through the development of natural characteristics which had lain dormant at first, or else they continue their lives, keeping time like two watches, hardly to be thrown out of accord except by some physical jar. Nature is far stronger than Nurture within the limited range that I have been careful to assign to the latter.

The effect of illness, as shown by these replies, is great, and well deserves further consideration. It appears that the constitution of youth is not so elastic as we are apt to think, but that an attack, say of scarlet fever, leaves a permanent mark, easily to be measured by the present method of comparison. This recalls an impression made strongly on my mind several years ago, by the sight of some curves drawn by a mathematical friend. He took monthly measurements of the circumference of his children's heads during the first few years of their lives, and he laid down the successive measurements on the successive lines of a piece of ruled paper, by taking the edge of the paper as a base. He then joined the free ends of the lines, and so obtained a curve of growth. These curves had, on the whole, that regularity of sweep that might have been expected, but each of them showed occasional halts, like the landing-places on a long flight of stairs. The development·had been arrested by something, and was not made up for by after growth. Now, on the same piece of paper my friend had also registered the various infantine illnesses of the children, and corresponding to each illness was one of these halts. There remained no doubt in my mind that, if these illnesses had been warded off, the development of the children would have been increased by almost the precise amount lost in these halts. In other words, the disease had drawn largely upon the capital, and not only on the income, of their constitutions. I hope these remarks may induce some men of science to repeat similar experiments on their children of the future. They may compress two years of a child's history on one side of a ruled half-sheet of foolscap paper, if they cause each successive line to stand for a successive month, beginning from the birth of the child; and if they economise space by laying, not the o-inch division of the tape against the edge of the pages, but, say, the 10-inch division.

The steady and pitiless march of the hidden weaknesses in our constitutions, through illness to death, is painfully revealed by these histories of twins. We are too apt to look upon illness and death as capricious events, and there are some who ascribe them to the direct effect of supernatural interference, whereas the fact of the maladies of two twins being continually alike shows that illness and death are necessary incidents in a regular sequence of constitutional changes beginning at birth, and upon which external circumstances have, on the whole, very small effect. In cases where the maladies of the twins are continually alike, the clocks of their two lives move regularly on at the same rate, governed by their internal mechanism. When the hands approach the hour, there are sudden clicks, followed by a whirring of wheels; the moment that they touch it, the strokes fall. Necessitarians may derive new arguments from the life-histories of twins.

We will now consider the converse side of our subject, which appears to me even the more important of the two. Hitherto we have investigated cases where the similarity at first was close, but afterwards became less; now we will examine those in which there was great dissimilarity at first, and will see how far an identity of nurture in childhood and youth tended to assimilate them. As has been already mentioned, there is a large proportion of cases of sharply-contrasted characteristics, both of body and mind, among twins   I have twenty such cases, given with much detail. It is a fact that extreme dissimilarity, such as existed between Esau and Jacob, is a no less marked peculiarity in twins of the same sex than extreme similarity.   On this curious point, and on much else in the history of twins, I have many remarks to make, but this is not the place to make them.

The evidence given by the twenty cases above mentioned is absolutely accordant, so that the character of the whole may be exactly conveyed by a few quotations.

(1.) One parent says:—"They have had *exactly the same nurture* from their birth up to the present time; they are both perfectly healthy and strong, yet they are otherwise as dissimilar as two boys could be, physically, mentally, and in their emotional nature."

(2.) "I can answer most decidedly that the twins have been perfectly dissimilar in character, habits, and likeness from the moment of their birth to the present time, though they were nursed by the same woman, went to school together, and were never separated till the age of fifteen."

(3.) "They have never been separated, never the least differently treated in food, clothing, or education; both teethed at the same time, both had measles, whooping-cough, and scarlatina at the same time, and neither had had any other serious illness.   Both are and have been exceedingly healthy, and have good abilities, yet they differ as much from each other in mental cast as any one of my family differs from another."

(4.) "Very dissimilar in body and mind: the one is quiet, retiring, and slow but sure; good-tempered, but disposed to be sulky when provoked;—the other is quick, vivacious, forward, acquiring easily and forgetting soon; quick-tempered and choleric, but quickly forgiving and forgetting.   They have been educated together and never separated."

(5.) "They were never alike either in body or mind, and their dissimilarity increases daily.   The external influences have been identical; they have never been separated."

(6.) "The two sisters are very different in ability and disposition.   The one is retiring, but firm and determined; she has no taste for music or drawing.   The other is of an active, excitable temperament: she displays an unusual amount of quickness and talent, and is passionately fond of music and drawing.   From infancy, they have been rarely separated even at school, and as children visiting their friends, they always went together."

(7.) "They have been treated exactly alike ; both were brought up by hand ; they have been under the same nurse and governess from their birth, and they are very fond of each other. Their increasing dissimilarity must be ascribed to a natural difference of mind and character, as there has been nothing in their treatment to account for it."

(8.) "They are as different as possible. [A minute and unsparing analysis of the characters of the two twins is given by their father, most instructive to read, but impossible to publish without the certainty of wounding the feelings of one of the twins, if these pages should chance to fall under his eyes.] They were brought up entirely by hand, that is, on cow's milk, and treated by one nurse in precisely the same manner."

(9.) "The home-training and influence were precisely the same, and therefore I consider the dissimilarity to be accounted for almost entirely by innate disposition and by causes over which we have no control."

(10.) "This case is, I should think, somewhat remarkable for dissimilarity in physique as well as for strong contrast in character. They have been unlike in body and mind throughout their lives. Both were reared in a country house, and both were at the same schools till æt. 16."

(11.) "Singularly unlike in body and mind from babyhood ; in looks, dispositions, and tastes they are quite different. I I think I may say the dissimilarity was innate, and developed more by time than circumstance."

(12.) "We were never in the least degree alike. I should say my sister's and my own character are diametrically opposed, and have been utterly different from our birth, though a very strong affection subsists between us."

(13.) The father remarks :—" They were curiously different in body and mind from their birth."

The surviving twin (a senior wrangler of Cambridge) adds :—
"A fact struck all our school contemporaries, that my brother and I were complementary, so to speak, in point of ability and disposition. He was contemplative, poetical, and literary to a remarkable degree, showing great power in that line. I was practical, mathematical, and linguistic. Between us we should have made a very decent sort of a man."

I could quote others just as strong as these, in some of which the above phrase "complementary" also appears, while I have not a single case in which my correspondents speak of originally dissimilar characters having become assimilated through identity of nurture. However, a somewhat exaggerated estimate of dissimilarity may be due to the tendency of relatives to dwell unconsciously on distinctive peculiarities, and to disregard the far more numerous points of likeness that would first attract the notice of a stranger. Thus in case 11 I find the remark, "Strangers see a strong likeness between them, but none who knows them well can perceive it." Instances are common of slight acquaintances mistaking members, and especially daughters of a family, for one another, between whom intimate friends can barely discover a resemblance. Still, making reasonable allowance for unintentional exaggeration, the impression that all this evidence leaves on the mind is one of some wonder whether nurture can do anything at all, beyond giving instruction and professional training. It emphatically corroborates and goes far beyond the conclusions to which we had already been driven by the cases of similarity. In those, the causes of divergence began to act about the period of adult life, when the characters had become somewhat fixed; but here the causes conducive to assimilation began to act from the earliest moment of the existence of the twins, when the disposition was most pliant, and they were continuous until the period of adult life. There is no escape from the conclusion that nature prevails enormously over nurture when the differences of nurture do not exceed what is commonly to be found among persons of the same rank of society and in the same country. My fear is, that my evidence may seem to prove too much, and be discredited on that account, as it appears contrary to all experience that nurture should go for so little. But experience is often fallacious in ascribing great effects to trifling circumstances. Many a person has amused himself with throwing bits of stick into a tiny brook and watching their progress; how they are arrested, first by one chance obstacle, then by another; and again, how their onward course is facilitated by a combination of circumstances. He might ascribe much importance to each of these events, and think how largely the destiny of the stick had been governed by a series of trifling accidents. Nevertheless all the sticks succeed in passing down the current, and in the long-run, they travel at nearly the same rate. So it is with life, in respect to the several accidents which seem to have had a great effect upon our careers. The one element, that varies in different individuals, but is constant in each of them, is the natural tendency; it corresponds to the current in the stream, and inevitably asserts itself.

Much stress is laid on the persistence of moral impressions made in childhood, and the conclusion is drawn, that the effects of early teaching must be important in a corresponding degree. I acknowledge the fact, so far as has been explained in the chapter on Early Sentiments, but there is a considerable set-off on the other side. Those teachings that conform to the natural aptitudes of the child leave much more enduring marks than others. Now both the teachings and the natural aptitudes of the child are usually derived from its parents. They are able to understand the ways of one another more intimately than is possible to persons not of the same blood, and the child instinctively assimilates the habits and ways of thought of its parents. Its disposition is "educated" by them, in the true sense of the word; that is to say, it is evoked, not formed by them. On these grounds I ascribe the persistence of many habits that date from early home education, to the peculiarities of the instructors rather than to the period when the instruction was given. The marks left on the memory by the instructions of a foster-mother are soon sponged clean away. Consider the history of the cuckoo, which is reared exclusively by foster-mothers. It is probable that nearly every young cuckoo, during a series of many hundred generations, has been brought up in a family whose language is a chirp and a twitter. But the cuckoo cannot or will not adopt that language, or any other of the habits of its foster-parents. It leaves its birthplace as soon as it is able, and finds out its own kith and kin, and identifies itself henceforth with them. So utterly are its earliest instructions in an alien bird-language neglected, and so completely is its new education successful, that the note of the cuckoo tribe is singularly correct.

DOMESTICATION OF ANIMALS.[1]

Before leaving the subject of Nature and Nurture, I would direct attention to evidence bearing on the conditions under

[1] This memoir is reprinted from the *Transactions of the Ethnological*

which animals appear first to have been domesticated. It clearly shows the small power of nurture against adverse natural tendencies.

The few animals that we now possess in a state of domestication were first reclaimed from wildness in prehistoric times. Our remote barbarian ancestors must be credited with having accomplished a very remarkable feat, which no subsequent generation has rivalled. The utmost that we of modern times have succeeded in doing, is to improve the races of those animals that we received from our forefathers in an already domesticated condition.

There are only two reasonable solutions of this exceedingly curious fact. The one is, that men of highly original ideas, like the mythical Prometheus, arose from time to time in the dawn of human progress, and left their respective marks on the world by being the first to subjugate the camel, the llama, the reindeer, the horse, the ox, the sheep, the hog, the dog, or some other animal to the service of man. The other hypothesis is that only a few species of animals are fitted by their nature to become domestic, and that these were discovered long ago through the exercise of no higher intelligence than is to be found among barbarous tribes of the present day. The failure of civilised man to add to the number of domesticated species would on this supposition be due to the fact that all the suitable material whence domestic animals could be derived has been long since worked out.

I submit that the latter hypothesis is the true one for the reasons about to be given; and if so, the finality of the process of domestication must be accepted as one of the most striking instances of the inflexibility of natural disposition, and of the limitations thereby imposed upon the

Society, 1865, with an alteration in the opening and concluding paragraphs, and with a few verbal emendations. If I had discussed the subject now for the first time I should have given extracts from the works of the travellers of the day, but it seemed needless to reopen the inquiry merely to give it a more modern air. I have also preferred to let the chapter stand as it was written, because considerable portions of it have been quoted by various authors (e.g. Bagehot, Economic Studies, pp. 161 to 166 : Longman, 1880), and the original memoir is not easily accessible.

choice of careers for animals, and by analogy for those of men.

My argument will be this:—All savages maintain pet animals, many tribes have sacred ones, and kings of ancient states have imported captive animals on a vast scale, for purposes of show, from neighbouring countries. I infer that every animal, of any pretensions, has been tamed over and over again, and has had numerous opportunities of becoming domesticated. But the cases are rare in which these opportunities have led to any result. No animal is fitted for domestication unless it fulfils certain stringent conditions, which I will endeavour to state and to discuss. My conclusion is, that all domesticable animals of any note have long ago fallen under the yoke of man. In short, that the animal creation has been pretty thoroughly, though half unconsciously, explored, by the every-day habits of rude races and simple civilisations.

It is a fact familiar to all travellers, that savages frequently capture young animals of various kinds, and rear them as favourites, and sell or present them as curiosities. Human nature is generally akin : savages may be brutal, but they are not on that account devoid of our taste for taming and caressing young animals ; nay, it is not improbable that some races may possess it in a more marked degree than ourselves, because it is a childish taste with us ; and the motives of an adult barbarian are very similar to those of a civilised child.

In proving this assertion, I feel embarrassed with the multiplicity of my facts. I have only space to submit a few typical instances, and must, therefore, beg it will be borne in mind that the following list could be largely reinforced. Yet even if I inserted all I have thus far been able to collect, I believe insufficient justice would be done to the real truth of the case. Captive animals do not commonly fall within the observation of travellers, who mostly confine themselves to their own encampments, and abstain from entering the dirty dwellings of the natives ; neither do the majority of travellers think tamed animals worthy of detailed mention. Consequently the anecdotes of their existence are scattered sparingly among a large number of volumes. It is when those travellers are questioned who have lived long and intimately with savage tribes that the plenitude of available instances becomes most apparent.

I proceed to give anecdotes of animals being tamed in various parts of the world, at dates when they were severally beyond the reach of civilised influences, and where, therefore, the pleasure taken by the natives in taming them must be ascribed to their unassisted mother-wit. It will be inferred that the same rude races who were observed to be capable of great fondness towards animals in particular instances, would not unfrequently show it in others.

[North America.]—The traveller Hearne, who wrote towards the end of the last century, relates the following story of moose or elks in the more northern parts of North America. He says :—

"I have repeatedly seen moose at Churchill as tame as sheep and even more so. . . . The same Indian that brought them to the Factory had, in the year 1770, two others so tame that when on his passage to Prince of Wales's Fort in a canoe, the moose always followed him along the bank of the river ; and at night, or on any other occasion when the Indians landed, the young moose generally came and fondled on them, as the most domestic animal would have done, and never offered to stray from the tents."

Sir John Richardson, in an obliging answer to my inquiries about the Indians of North America, after mentioning the bison calves, wolves, and other animals that they frequently capture and keep, said :—

"It is not unusual, I have heard, for the Indians to bring up young bears, the women giving them milk from their own breasts."

He mentions that he himself purchased a young bear, and adds :—

"The red races are fond of pets and treat them kindly ; and in purchasing them there is always the unwillingness of the women and children to overcome, rather than any dispute about price. My young bear used to rob the women of the berries they had gathered, but the loss was borne with good nature."

I will again quote Hearne, who is unsurpassed for his

minute and accurate narratives of social scenes among the Indians and Esquimaux.   In speaking of wolves he says :—

"They always burrow underground to bring forth their young, and though it is natural to suppose them very fierce at those times, yet I have frequently seen the Indians go to their dens and take out the young ones and play with them.   I never knew a Northern Indian hurt one of them ; on the contrary, they always put them carefully into the den again ; and I have sometimes seen them paint the faces of the young wolves with vermilion or red ochre."

[South America.]—Ulloa, an ancient traveller, says :—

"Though the Indian women breed fowl and other domestic animals in their cottages, they never eat them : and even conceive such a fondness for them, that they will not sell them, much less kill them with their own hands.   So that if a stranger who is obliged to pass the night in one of their cottages, offers ever so much money for a fowl, they refuse to part with it, and he finds himself under the necessity of killing the fowl himself. At this his landlady shrieks, dissolves into tears, and wrings her hands, as if it had been an only son, till seeing the mischief past mending, she wipes her eyes and quietly takes what the traveller offers her."

The care of the South American Indians, as Quiloa truly states, is by no means confined to fowls.   Mr. Bates, the distinguished traveller and naturalist of the Amazons, has favoured me with a list of twenty-two species of quadrupeds that he has found tame in the encampments of the tribes of that valley.   It includes the tapir, the agouti, the guinea-pig, and the peccari.   He has also noted five species of quadrupeds that were in captivity, but not tamed.   These include the jaguar, the great ant-eater, and the armadillo.   His list of tamed birds is still more extensive.

[North Africa.]—The ancient Egyptians had a positive passion for tamed animals, such as antelopes, monkeys, crocodiles, panthers, and hyenas.   Mr. Goodwin, the eminent Egyptologist, informed me that "they anticipated our zoological tastes completely," and that some of the pictures referring to tamed animals are among their very earliest monuments, viz. 2000 or 3000 years B.C.   Mr. Mansfield Parkyns, who passed many years in Abyssinia

N

and the countries of the Upper Nile, writes me word in answer to my inquiries :—

"I am sure that negroes often capture and keep alive wild animals. I have bought them and received them as presents— wild cats, jackals, panthers, the wild dog, the two best lions now in the Zoological Gardens, monkeys innumerable and of all sorts, and mongoos. I cannot say that I distinctly recollect any pets among the *lowest* orders of men that I met with, such as the Denkas, but I am sure they exist, and in this way. When I was on the White Nile and at Khartoum, very few merchants went up the White Nile ; none had stations. They were little known to the natives ; but none returned without some live animal or bird which they had procured from them. While I was at Khartoum, there came an Italian wild beast showman, after the Wombwell style. He made a tour of the towns up to Doul and Fazogly, Kordofan and the peninsula, and collected a large number of animals. Thus my opinion distinctly is, that negroes do keep wild animals alive. *I am sure of it;* though I can only vaguely recollect them in one or two cases. I remember some chief in Abyssinia who had a pet lion which he used to tease, and I have often seen monkeys about huts."

[Equatorial Africa.]—The most remarkable instance I have met with in modern Africa is the account of a menagerie that existed up to the beginning of the reign of the present king of the Wahumas, on the shores of Lake Nyanza. Suna, the great despot of that country, reigned till 1857. Captains Burton and Speke were in the neighbourhood in the following year, and Captain Burton thus describes (*Journal R. G. Soc.*, xxix. 282) the report he received of Suna's collection :—

"He had a large menagerie of lions, elephants, leopards, and similar beasts of disport ; he also kept for amusement fifteen or sixteen albinos ; and so greedy was he of novelty, that even a cock of peculiar form or colour would have been forwarded by its owner to feed his eyes."

Captain Speke, in his subsequent journey to the Nile, passed many months at Uganda, as the guest of Suna's youthful successor, M'tese. The fame of the old menagerie was fresh when Captain Speke was there. He wrote to me as follows concerning it :—

"I was told Suna kept buffaloes, antelopes, and animals of all colours' (meaning 'sorts'), and in equal quantities. M'tese, his son, no sooner came to the throne, than he indulged in shooting them down before his admiring wives, and now he has only one buffalo and a few parrots left."

In Kouka, near Lake Tchad, antelopes and ostriches are both kept tame, as I was informed by Dr. Barth.

[South Africa.]—The instances are very numerous in South Africa where the Boers and half-castes amuse themselves with rearing zebras, antelopes, and the like; but I have not found many instances among the native races. Those that are best known to us are mostly nomad and in a chronic state of hunger, and therefore disinclined to nurture captured animals as pets; nevertheless, some instances can be adduced. Livingstone alludes to an extreme fondness for small tame singing-birds (pp. 324 and 453). Dr. (now Sir John) Kirk, who accompanied him in later years, mentions guinea-fowl—that do not breed in confinement, and are merely kept as pets—in the Shiré valley, and Mr. Oswell has furnished me with one similar anecdote. I feel, however, satisfied that abundant instances could be found if properly sought for. It was the frequency with which I recollect to have heard of tamed animals when I myself was in South Africa, though I never witnessed any instance, that first suggested to me the arguments of the present paper. Sir John Kirk informs me that:

"As you approach the coast or Portuguese settlements, pets of all kinds become very common; but then the opportunity of occasionally selling them to advantage may help to increase the number; still, the more settled life has much to do with it."

In confirmation of this view, I will quote an early writer, Pigafetta (*Hakluyt Coll.*, ii. 562), on the South African kingdom of Congo, who found a strange medley of animals in captivity, long before the demands of semi-civilisation had begun to prompt their collection:—

The King of Congo, on being Christianised by the Jesuit missionaries in the sixteenth century, "signified that whoever had any idols should deliver them to the lieutenants of the country. And within less than a month all the idols which they worshipped were brought into court, and certainly the number of these toys was infinite, for every man adored what he liked without any measure or reason at all. Some kept serpents of horrible figures, some worshipped the greatest goats they could get, some leopards, and others monstrous creatures. Some held in veneration certain unclean fowls, etc. Neither did they content themselves with worshipping the said creatures when alive, but also adored the very skins of them when they were dead and stuffed with straw."

[Australia.]—Mr. Woodfield records the following touching anecdote in a paper communicated to the Ethnological Society, as occurring in an unsettled part of West Australia, where the natives rank as the lowest race upon the earth :—

"During the summer of 1858-9 the Murchison river was visited by great numbers of kites, the native country of these birds being Shark's Bay. As other birds were scarce, we shot many of these kites, merely for the sake of practice, the natives eagerly devouring them as fast as they were killed. One day a man and woman, natives of Shark's Bay, came to the Murchison, and the woman immediately recognising the birds as coming from her country, assured us that the natives there never kill them, and that they are so tame that they will perch on the shoulders of the women and eat from their hands. On seeing one shot she wept bitterly, and not even the offer of the bird could assuage her grief, for she absolutely refused to eat it. No more kites were shot while she remained among us."

The Australian women habitually feed the puppies they intend to rear from their own breasts, and show an affection to them equal, if not exceeding, that to their own infants. Sir Charles Nicholson informs me that he has known an extraordinary passion for cats to be demonstrated by Australian women at Fort Phillip.

[New Guinea Group.]—Captain Develyn is reported (Bennett, *Naturalist in Australia*, p. 244) to say of the island of New Britain, near Australia, that the natives consider cassowaries " to a certain degree sacred, and rear them as pets. They carry them in their arms, and entertain a great affection for them."

Professor Huxley informs me that he has seen sucking-pigs nursed at the breasts of women, apparently as pets, in islands of the New Guinea Group.

[Polynesia.]—The savage and cannibal Fijians were no exceptions to the general rule, for Dr. Seemann wrote me word that they make pets of the flying fox (bat), the lizard, and parroquet. Captain Wilkes, in his exploring expedition (ii. 122), says the pigeon in the Samoon islands "is commonly kept as a plaything, and particularly by the chiefs. One of our officers unfortunately on one occasion shot a pigeon, which caused great commotion, for the bird was a king pigeon, and to kill it was thought as great a crime as to take the life of a man."

Mr. Ellis, writing of these islands (*Polynesian Researches*, ii. 285), says :—

" Eels are great favourites, and are tamed and fed till they attain an enormous size. Taoarii had several in different parts of the island. These pets were kept in large holes, two or three feet deep, partially filled with water. I have been several times with the young chief, when he has sat down by the side of the hole, and by giving a shrill sort of whistle, has brought out an enormous eel, which has moved about the surface of the water and eaten with confidence out of his master's hand."

[Syria.]—I will conclude this branch of my argument by quoting the most ancient allusion to a pet that I can discover in writing, though some of the Egyptian pictured representations are considerably older. It is the parable spoken by the Prophet Samuel to King David, that is expressed in the following words :—

" The poor man had nothing save one little ewe lamb, which he had bought and nourished up : and it grew up together with him and with his children; it did eat of his own meat, and drank of his own cup, and lay in his bosom, and was to him as a daughter."

We will now turn to the next stage of our argument. Not only do savages rear animals as pets, but communities maintain them as sacred. The ox of India and the brute gods of Egypt occur to us at once; the same superstition prevails widely. The quotation already given from Pigafetta is in point ; the fact is too well known to readers of travel to make it necessary to devote space to its proof. I will therefore simply give a graphic account, written by M. Jules Gérard, of Whydah in West Africa :—

" I visited the Temple of Serpents in this town, where thirty of these monstrous deities were asleep in various attitudes. Each day at sunset, a priest brings them a certain number of sheep, goats, fowls, etc., which are slaughtered in the temple and then divided among the 'gods.' Subsequently during the night they (? the priests) spread themselves about the town, entering the houses in various quarters in search of further offerings. It is forbidden under penalty of death to kill, wound, or even strike one of these sacred serpents, or any other of the same species, and only the priests possess the privilege of taking hold of them, for the purpose of reinstating them in the temple should they be found elsewhere."

It would be tedious and unnecessary to adduce more instances of wild animals being nurtured in the encampments of savages, either as pets or as sacred animals. It will be found on inquiry that few travellers have failed altogether to observe them. If we consider the small number of encampments they severally visited in their line of march, compared with the vast number that are spread over the whole area, which is or has been inhabited by rude races, we may obtain some idea of the thousands of places at which half-unconscious attempts at domestication are being made in each year. These thousands must themselves be multiplied many thousandfold, if we endeavour to calculate the number of similar attempts that have been made since men like ourselves began to inhabit the world.

My argument, strong as it is, admits of being considerably strengthened by the following consideration :—

The natural inclination of barbarians is often powerfully reinforced by an enormous demand for captured live animals on the part of their more civilised neighbours. A desire to create vast hunting-grounds and menageries and amphitheatrical shows, seems naturally to occur to the monarchs who preside over early civilisations, and travellers continually remark that, whenever there is a market for live animals, savages will supply them in any quantities. The means they employ to catch game for their daily food readily admits of their taking them alive. Pit-falls, stake-nets, and springes do not kill. If the savage captures an animal unhurt, and can make more by selling it alive than dead, he will doubtless do so. He is well fitted by education to keep a wild animal in captivity. His mode of pursuing game requires a more intimate knowledge of the habits of beasts than is ever acquired by sportsmen who use more perfect weapons. A savage is obliged to steal upon his game, and to watch like a jackal for the leavings of large beasts of prey. His own mode of life is akin to that of the creatures he hunts. Consequently, the savage is a good gamekeeper; captured animals thrive in his charge, and he finds it remunerative to take them a long way to market. The demands of ancient Rome appear to have penetrated Northern Africa as far or farther than the steps of our modern explorers. The chief centres of import of wild animals were Egypt, Assyria (and other Eastern monarchies), Rome, Mexico, and Peru. I have not yet been able to learn what were the habits of Hindostan or China. The modern menagerie of Lucknow is the only considerable native effort in those parts with which I am acquainted.

[Egypt.]—The mutilated statistical tablet of Karnak (*Trans. R. Soc. Lit.*, 1847, p. 369, and 1863, p. 65) refers to an armed invasion of Armenia by Thothmes III., and the payment of a large tribute of antelopes and birds. When Ptolemy Philadelphus fêted the Alexandrians (*Athenæus*, v.), the Ethiopians brought dogs, buffaloes, bears, leopards, lynxes, a giraffe, and a rhinoceros. Doubtless this description of gifts was common. Live beasts are the one article of curiosity and amusement that barbarians can offer to civilised nations.

[Assyria.]—Mr. Fox Talbot thus translates (*Journal Asiatic Soc.*, xix. 124) part of the inscription on the black obelisk of Ashurakbal found in Nineveh and now in the British Museum :—

"He caught in hunter's toils (a blank number) of armi, turakhi, nali, and yadi. Every one of these animals he placed in separate enclosures. He brought up their young ones and counted them as carefully as young lambs. As to the creatures called burkish, utrati (dromedaries ?), tishani, and dagari, he wrote for them and they came. The dromedaries he kept in enclosures, where he brought up their young ones. He entrusted each kind of animal to men of their own country to tend them. There were also curious animals of the Mediterranean Sea, which the King of Egypt sent as a gift and entrusted to the care of men of their own land. The very choicest animals were there in abundance, and birds of heaven with beautiful wings. It was a splendid menagerie, and all the work of his own hands. The names of the animals were placed beside them."

[Rome.]—The extravagant demands for the amphitheatre of ancient Rome must have stimulated the capture of wild animals in Asia, Africa, and the then wild parts of Europe, to an extraordinary extent. I will quote one instance from Gibbon :—

" By the order of Probus, a vast quantity of large trees torn up by the roots were transplanted into the midst of the circus. The spacious and shady forest was immediately filled with a thousand ostriches, a thousand stags, a thousand fallow-deer, and a thousand wild boars, and all this variety of game was abandoned to the riotous impetuosity of the multitude. The tragedy of the succeeding day consisted in the massacre of a hundred lions, an equal number of lionesses, two hundred leopards, and three hundred bears."

Farther on we read of a spectacle by the younger Gordian of "twenty zebras, ten elks, ten giraffes, thirty African hyenas, ten Indian tigers, a rhinoceros, an hippopotamus, and thirty-two elephants."

[Mexico.]—Gomara, the friend and executor of Herman Cortes, states :—

" There were here also many cages made of stout beams, in some of which there were lions (pumas) ; in others, tigers (jaguars) ; in others, ounces ; in others, wolves ; nor was there any animal on four legs that was not there. They had for their rations deer and other animals of the chase. There were also kept in large jars or tanks, snakes, alligators, and lizards. In another court there were cages containing every kind of birds of prey, such as vultures, a dozen sorts of falcons and hawks, eagles, and owls. The large eagles received turkeys for their food. Our Spaniards were astonished at seeing such a diversity of birds and beasts ; nor did they find it pleasant to hear the hissing of the poisonous snakes, the roaring of the lions, the shrill cries of the wolves, nor the groans of the other animals given to them for food."

[Peru.]—Garcilasso de la Vega (*Commentarios Reales*, v. 10), the son of a Spanish conqueror by an Indian princess, born and bred in Peru, writes :—

" All the strange birds and beasts which the chiefs presented to the Inca were kept at court, both for grandeur and also to please the Indians who presented them. When I came to Cuzco, I remember there were some remains of places where they kept these creatures. One was the serpent conservatory, and another where they kept the pumas, jaguars, and bears."

[Syria and Greece.]—I could have said something on Solomon's apes and peacocks, and could have quoted at length the magnificent order given by Alexander the Great (Pliny, *Nat. Hist.*, viii. 16) towards supplying material for Aristotle's studies in natural history ; but enough has been said to prove what I maintained, namely, that numerous cases occur, year after year, and age after age, in which every animal of note is captured and its capabilities of domestication unconsciously tested.

I would accept in a more stringent sense than it was probably intended to bear, the text of St. James, who wrote at a time when a vast variety and multitude of animals were constantly being forwarded to Rome and to Antioch for amphitheatrical shows. He says (James iii. 7), " Every kind of beasts, and of birds, and of serpents, and of things in the sea, is tamed, and hath been tamed of mankind."

I conclude from what I have stated that there is no animal worthy of domestication that has not frequently been captured, and might ages ago have established itself as a domestic breed, had it not been deficient in certain necessary particulars which I shall proceed to discuss. These are numerous and so stringent as to leave no ground for wonder that out of the vast abundance of the animal creation, only a few varieties of a few species should have become the companions of man.

It by no means follows that because a savage cares to take home a young fawn to amuse himself, his family, and his friends, that he will always continue to feed or to look after it. Such attention would require a steadiness of purpose foreign to the ordinary character of a savage. But herein lie two shrewd tests of the eventual destiny of the animal as a domestic species.

Hardiness.—It must be able to shift for itself and to thrive, although it is neglected ; since, if it wanted much care, it would never be worth its keep.

The hardiness of our domestic animals is shown by the

rapidity with which they establish themselves in new lands. The goats and hogs left on islands by the earlier navigators throve excellently on the whole. The horse has taken possession of the Pampas, and the sheep and ox of Australia. The dog is hardly repressible in the streets of an Oriental town.

Fondness of Man.—Secondly, it must cling to man, notwithstanding occasional hard usage and frequent neglect. If the animal had no natural attachment to our species, it would fret itself to death, or escape and revert to wildness. It is easy to find cases where the partial or total non-fulfilment of this condition is a corresponding obstacle to domestication. Some kinds of cattle are too precious to be discarded, but very troublesome to look after. Such are the reindeer to the Lapps. Mr. Campbell of Islay informed me that the tamest of certain herds of them look as if they were wild ; they have to be caught with a lasso to be milked. If they take fright, they are off to the hills ; consequently the Lapps are forced to accommodate themselves to the habits of their beasts, and to follow them from snow to sea and from sea to snow at different seasons. The North American reindeer has never been domesticated, owing, I presume, to this cause. The Peruvian herdsmen would have had great trouble to endure had the llama and alpaca not existed, for their cogeners, the huanacu and the vicuna, are hardly to be domesticated.

Zebras, speaking broadly, are unmanageable. The Dutch Boers constantly endeavour to break them to harness, and though they occasionally succeed to a degree, the wild mulish nature of the animal is always breaking out, and liable to balk them.

It is certain that some animals have naturally a greater fondness for man than others ; and as a proof of this, I will again quote Hearne about the moose, who are considered by him to be the easiest to tame and domesticate of any of the deer tribe. Formerly the closely-allied European elks were domesticated in Sweden, and used to draw sledges, as they are now occasionally in Canada ; but they have been obsolete for many years. Hearne says :—

" The young ones are so simple that I remember to have seen an Indian paddle his canoe up to one of them, and take it by the poll, without experiencing the least opposition, the poor harmless animal seeming at the same time as contented alongside the canoe as if swimming by the side of its dam, and looking up in our faces with the same fearless innocence that a house lamb would."

On the other hand, a young bison will try to dash out its brains against the tree to which it is tied, in terror and hatred of its captors.

It is interesting to note the causes that conduce to a decided attachment of certain animals to man, or between one kind of animal and another. It is notorious that attachments and aversions exist in nature. Swallows, rooks, and storks frequent dwelling houses; ostriches and zebras herd together; so do bisons and elks. On the other hand, deer and sheep, which are both gregarious, and both eat the same food and graze within the same enclosure, avoid one another. The spotted Danish dog, the Spitz dog, and the cat, have all a strong attachment to horses, and horses seem pleased with their company; but dogs and cats are proverbially discordant. I presume that two species of animals do not consider one another companionable, or clubable, unless their behaviour and their persons are reciprocally agreeable. A phlegmatic animal would be exceedingly disquieted by the close companionship of an excitable one. The movements of one beast may have a character that is unpleasing to the eyes of another; his cries may sound discordant; his smell may be repulsive. Two herds of animals would hardly intermingle, unless their respective languages of action and of voice were mutually intelligible. The animal which above all others is a companion to man is the dog, and we observe how readily their proceedings are intelligible to each other. Every whine or bark of the dog, each of his fawning, savage, or timorous movements is the exact counterpart of what would have been the man's behaviour, had he felt similar emotions. As the man understands the thoughts of the dog, so the dog understands the thoughts of the man, by attending to his natural voice, his countenance, and his actions. A man irritates a dog by an ordinary laugh, he frightens him by an angry look, or he calms him by a kindly bearing; but he has less spontaneous hold over an ox or a sheep. He must study their ways and tutor his behaviour before he can either understand the feelings of those animals or make his own intelligible to them. He has no natural power at all over many other creatures. Who, for instance, ever succeeded in frowning away a mosquito, or in pacifying an angry wasp by a smile?

Desire of Comfort.—This is a motive which strongly attaches certain animals to human habitations, even though they are unwelcome : it is a motive which few persons who have not had an opportunity of studying animals in savage lands are likely to estimate at its true value. The life of all beasts in their wild state is an exceedingly anxious one. From my own recollection, I believe that every antelope in South Africa has to run for its life every one or two days upon an average, and that he starts' or gallops under the influence of a false alarm many times in a day. Those who have crouched at night by the side of pools in the desert, in order to have a shot at the beasts that frequent them, see strange scenes of animal life ; how the creatures gambol at one moment and fight at another; how a herd suddenly halts in strained attention, and then breaks into a maddened rush, as one of them becomes conscious of the stealthy movements or rank scent of a beast of prey. Now this hourly life-and-death excitement is a keen delight to most wild creatures, but must be peculiarly distracting to the comfort-loving temperament of others. The latter are alone suited to endure the crass habits and dull routine of domesticated life. Suppose that an animal which has been captured and half-tamed, received ill-usage from his captors, either as punishment or through mere brutality, and that he rushed indignantly into the forest with his ribs aching from blows and stones. If a comfort-loving animal, he will probably be no gainer by the change, more serious alarms and no less ill-usage awaits him ; he hears the roar of the wild beasts and the headlong gallop of the frightened herds, and he finds the buttings and the kicks of other animals harder to endure than the blows from which he fled. He has the disadvantage of being a stranger, for the herds of his own species which he seeks for companionship constitute so many cliques, into which he can only find admission by more fighting with their strongest members than he has spirit to undergo. As a set-off against these miseries, the freedom of savage life has no charms for his temperament; so the end of it is, that with a heavy heart he turns back to the habitation he had quitted. When animals thoroughly enjoy the excitement of wild life, I presume they cannot be domesticated, they could only be tamed, for they would never return from the joys of the wilderness after they had once tasted them through some accidental wandering.

Gallinas, or guinea-fowl, have so little care for comfort, or indeed for man, that they fall but a short way within the frontier of domestication. It is only in inclement seasons that they take contentedly to the poultry-yards.

Elephants, from their size and power, are not dependent on man for protection; hence, those that have been reared as pets from the time they were calves, and have never learned to dread and obey the orders of a driver, are peculiarly apt to revert to wildness if they once are allowed to wander and escape to the woods. I believe this tendency, together with the cost of maintenance and the comparative uselessness of the beasts, are among the chief causes why Africans never tame them now; though they have not wholly lost the practice of capturing them when full-grown, and of keeping them imprisoned for some days alive. Mr. Winwood Reade's account of captured elephants, seen by himself near Glass Town in Equatorial Western Africa, is very curious.

Usefulness to Man.—To proceed with the list of requirements which a captured animal must satisfy before it is possible he could be permanently domesticated : there is the very obvious condition that he should be useful to man; otherwise, in growing to maturity, and losing the pleasing youthful ways which had first attracted his captors and caused them to make a pet of him, he would be repelled. As an instance in point, I will mention seals. Many years ago I used to visit Shetland, when those animals were still common, and I heard many stories of their being tamed : one will suffice :—A fisherman caught a young seal; it was very affectionate, and frequented his hut, fishing for itself in the sea. At length it grew self-willed and unwieldy; it used to push the children and snap at strangers, and it was voted a nuisance, but the people could not bear to kill it on account of its human ways. One day the fisherman took it with him in his boat, and dropped it in a stormy sea, far from home; the stratagem was unsuccessful; in a day or two the well-known scuffling sound of the seal, as it floundered up to the hut, was again heard; the animal had found its way home. Some days after the poor creature was shot by a sporting stranger, who saw it basking and did not know it was tame. Now had the seal been a useful animal and not troublesome, the fisherman would doubtless have caught others, and set a watch over them to protect them; and then, if they bred freely and were easy to tend, it is likely enough he would have produced a domestic breed.

The utility of the animals as a store of future food is undoubtedly the most durable reason for maintaining them; but I think it was probably not so early a motive as the chief's pleasure in possessing them. That was the feeling under which the menageries, described above, were established. Whatever the despot of savage tribes is pleased with becomes invested with a sort of sacredness. His tame animals would be the care of all his people, who would become skilful herdsmen under the pressure of fear. It would be as much as their lives were worth if one of the creatures were injured through their neglect. I believe that the keeping of a herd of beasts, with the sole motive of using them as a reserve for food, or as a means of barter, is a late idea in the history of civilisation. It has now become established among the pastoral races of South Africa, owing to the traffickings of the cattle-traders, but it was by no means prevalent in Damara-Land when I travelled there in 1852. I then was surprised to observe the considerations that induced the chiefs to take pleasure in their vast herds of cattle. They were valued for their stateliness and colour, far more than for their beef. They were as the deer of an English squire, or as the stud of a man who has many more horses than he can ride. An ox was almost a sacred beast in Damara-Land, not to be killed except on momentous occasions, and then as a sort of sacrificial feast, in which all bystanders shared. The payment of two oxen was hush-money for the life of a man. I was considerably embarrassed by finding that I had the greatest trouble in buying oxen for my own use, with the ordinary articles of barter. The possessor would hardly part with them for any remuneration ; they would never sell their handsomest beasts.

One of the ways in which the value of tamed beasts would be soon appreciated would be that of giving milk to children. It is marvellous how soon goats find out children and tempt them to suckle. I have had the milk of my goats, when encamping for the night in African travels, drained dry by small black children, who had not the strength to do more than crawl about, but nevertheless came to some secret understanding with the goats and fed themselves. The records of many nations have legends like that of Romulus and Remus, who are stated to have been suckled by wild beasts. These are surprisingly confirmed by General Sleeman's narrative of six cases where children were nurtured for many years by wolves in Oude. (*Journey through Oude in 1849–50*, i. 206.)

Breeding freely.—Domestic animals must breed freely under confinement. This necessity limits very narrowly the number of species which might otherwise have been domesticated. It is one of the most important of all the conditions that have to be satisfied. The North American turkey, reared from the eggs of the wild bird, is stated to be unknown in the third generation, in captivity. Our turkey comes from Mexico, and was abundantly domesticated by the ancient Mexicans.

The Indians of the Upper Amazon took turtle and placed them in lagoons for use in seasons of scarcity. The Spaniards who first saw them called these turtle "Indian cattle." They would certainly have become domesticated like cattle, if they had been able to breed in captivity.

Easy to tend.—They must be tended easily. When animals reared in the house are suffered to run about in the companionship of others like themselves, they naturally revert to much of their original wildness. It is therefore essential to domestication that they should possess some quality by which large numbers of them may be controlled by a few herdsmen. The instinct of gregariousness is such a quality. The herdsman of a vast troop of oxen grazing in a forest, so long as he is able to see one of them, knows pretty surely that they are all within reach. If oxen are frightened and gallop off, they do not scatter, but remain in a single body. When animals are not gregarious, they are to the herdsman like a falling necklace of beads whose string is broken, or as a handful of water escaping between the fingers.

The cat is the only non-gregarious domestic animal. It is retained by its extraordinary adhesion to the comforts of the house in which it is reared.

An animal may be perfectly fitted to be a domestic animal, and be peculiarly easy to tend in a general way, and yet the circumstances in which the savages are living may make it too troublesome for them to maintain a breed. The following account, taken from Mr. Scott Nind's paper on the Natives of King George's Sound in Australia, and printed in the first volume of the *Journal of the Geographical Society*, is particularly to the point. He says:—

"In the chase the hunters are assisted by dogs, which they take when young and domesticate ; but they take little pains to train them to any particular mode of hunting. After finding a litter of young, the natives generally carry away one or two to rear ; in this case, it often occurs that the mother will trace and attack them ; and, being large and very strong, she is rather formidable. At some periods, food is so scanty as to compel the dog to leave his master and provide for himself; but in a few days he generally returns."

I have also evidence that this custom is common to the wild natives of other parts of Australia.

The gregariousness of all our domestic species is, I think, the primary reason why some of them are extinct in a wild state. The wild herds would intermingle with the tame ones, some would become absorbed, the others would be killed by hunters, who used the tame cattle as a shelter to approach the wild. Besides this, comfort-loving animals would be less suited to fight the battle of life with the rest of the brute creation ; and it is therefore to be expected that those varieties which are best fitted for domestication, would be the soonest extinguished in a wild state. For instance, we could hardly fancy the camel to endure in a land where there were large wild beasts.

Selection.—The irreclaimably wild members of every flock would escape and be utterly lost ; the wilder of those that remained would assuredly be selected for slaughter, whenever it was necessary that one of the flock should be killed. The tamest cattle—those that seldom ran away, that kept the flock together and led them homewards—would be preserved alive longer than any of the others. It is therefore these that chiefly become the parents of stock, and bequeath their domestic aptitudes to the future herd. I have constantly witnessed this process of selection among the pastoral savages of South Africa. I believe it to be a very important one, on account of its rigour and its regularity. It must have existed from the earliest times, and have been in continuous operation, generation after generation, down to the present day.

Exceptions.—I have already mentioned the African elephant, the North American reindeer, and the apparent, but not real exception of the North American turkey. I should add the ducks and geese of North America, but I cannot consider them in the light of a very strong case, for a savage who constantly changes his home is not likely to carry aquatic birds along with him. Beyond these few, I know of no notable exceptions to my theory.

Summary.

I see no reason to suppose that the first domestication of any animal, except the elephant, implies a high civilisation among the people who established it. I cannot believe it to have been the result of a preconceived intention, followed by elaborate trials, to administer to the comfort of man. Neither can I think it arose from one successful effort made by an individual, who might thereby justly claim the title of benefactor to his race ; but, on the contrary, that a vast number of half-unconscious attempts have been made throughout the course of ages, and that ultimately, by slow degrees, after many relapses, and continued selection, our several domestic breeds became firmly established.

I will briefly restate what appear to be the conditions under which wild animals may become domesticated :—1, they should be hardy ; 2, they should have an inborn liking for man ; 3, they should be comfort-loving ; 4, they should be found useful to the savages ; 5, they should breed freely ; 6, they should be easy to tend.

O

It would appear that every wild animal has had its chance of being domesticated, that those few which fulfilled the above conditions were domesticated long ago, but that the large remainder, who fail sometimes in only one small particular, are destined to perpetual wildness so long as their race continues. As civilisation extends they are doomed to be gradually destroyed off the face of the earth as useless consumers of cultivated produce. I infer that slight differences in natural dispositions of human races may in one case lead irresistibly to some particular career, and in another case may make that career an impossibility.

The Observed Order of Events.

There is nothing as yet observed in the order of events to make us doubt that the universe is bound together in space and time, as a single entity, and there is a concurrence of many observed facts to induce us to accept that view. We may, therefore, not unreasonably profess faith in a common and mysterious whole, and of the laborious advance, under many restrictions, of that infinitely small part of it which falls under our observation, but which is in itself enormously large, and behind which lies the awful mystery of the origin of all existence.

The conditions that direct the order of the whole of the living world around us, are marked by their persistence in improving the birthright of successive generations. They determine, at much cost of individual comfort, that each plant and animal shall, on the general average, be endowed at its birth with more suitable natural faculties than those of its representative in the preceding generation. They ensure, in short, that the inborn qualities of the terrestrial tenantry shall become steadily better adapted to their homes and to their mutual needs. This effect, be it understood, is not only favourable to the animals who live long enough to become parents, but is also favourable to those who perish in earlier life, because even they are on the whole better off during their brief career than if they had been born still less adapted to the conditions of their existence. If we summon before our imagination in a single mighty host, the whole number of living things from the earliest date at which terrestrial life can be deemed to have probably existed, to the latest future at which we may think it can probably continue, and if we cease to dwell on the miscarriages of individual lives or of single generations, we shall plainly perceive that the actual tenantry of the world progresses in a direction that may in some sense be described as the greatest happiness of the greatest number.

We also remark that while the motives by which individuals in the lowest stages are influenced are purely self regarding, they broaden as evolution goes on. The word " self " ceases to be wholly personal, and begins to include subjects of affection and interest, and these become increasingly numerous as intelligence and depth of character develop, and as civilisation extends. The sacrifice of the personal desire for repose to the performance of domestic and social duties is an everyday event with us, and other sacrifices of the smaller to the larger self are by no means uncommon. Life in general may be looked upon as a republic where the individuals are for the most part unconscious that while they are working for themselves they are also working for the public good.

We may freely confess ignorance of the outcome in the far future of that personal life to which we each cling passionately in the joyous morning of the affections, but which, as these and other interests fail, does not seem so eminently desirable in itself. We know that organic life can hardly be expected to flourish on this earth of ours for so long a time as it has already existed, because the sun will in all probability have lost too much of its heat and light by then, and will have begun to grow dark and therefore cold, as other stars have done. The conditions of existence here, which are now apparently in their prime, will have become rigorous and increasingly so, and there will be retrogression towards lower types, until the simplest form of life shall have wholly disappeared from the ice-bound surface. The whole living world will then have waxed and waned like an individual life.

Neither can we discover whether organisms here are capable of attaining the average development of organisms in other of the planets that are probably circling round most of the myriads of stars, whose physical constitution, where-ever it has as yet been observed spectroscopically, does not differ much from that of our sun. But we perceive around us a countless number of abortive seeds and germs; we find out of any group of a thousand men selected at random, some who are crippled, insane, idiotic, and otherwise born incurably imperfect in body or mind, and it is possible that this world may rank among other worlds as one of these.

We as yet understand nothing of the way in which our conscious selves are related to the separate lives of the billions of cells of which the body of each of us is composed. We only know that the cells form a vast nation, some members of which are always dying and others growing to supply their places, and that the continual sequence of these multitudes of little lives has its outcome in the larger and conscious life of the man as a whole. Our part in the universe may possibly in some distant way be analogous to that of the cells in an organised body, and our personalities may be the transient but essential elements of an immortal and cosmic mind.

Our views of the object of life have to be framed so as not to be inconsistent with the observed facts from which these various possibilities are inferred ; it is safer that they should not exclude the possibilities themselves. We must look on the slow progress of the order of evolution, and the system of routine by which it has thus far advanced, as due to antecedents and to inherent conditions of which we have not as yet the slightest conception. It is difficult to with-stand a suspicion that the three dimensions of space and the fourth dimension of time may be four independent variables of a system that is neither space nor time, but something else wholly unconceived by us. Our present enigma as to how a First Cause could itself have been brought into existence—how the tortoise of the fable, that bears the elephant that bears the world, is itself supported, —may be wholly due to our necessary mistranslation of the four or more variables of the universe, limited by inherent conditions, into the three unlimited variables of Space and the one of Time.

Our ignorance of the goal and purport of human life, and the mistrust we are apt to feel of the guidance of the spiritual sense, on account of its proved readiness to accept illusions as realities, warn us against deductive theories of conduct. Putting these, then, at least for the moment, to one side, we find ourselves face to face with two great and indisputable facts that everywhere force themselves on the attention and compel consideration. The one is that the whole of the living world moves steadily and continuously towards the evolution of races that are progressively more and more adapted to their complicated mutual needs and to their external circumstances. The other is that the process of evolution has been hitherto apparently carried out with, what we should reckon in our ways of carrying out projects, great waste of opportunity and of life, and with little if any consideration for individual mischance. Measured by our criterion of intelligence and mercy, which consists in the achievement of result without waste of time or opportunity, without unnecessary pain, and with equitable allowance for pure mistake, the process of evolution on this earth, so far as we can judge, has been carried out neither with intelligence nor ruth, but entirely through the routine of various sequences, commonly called "laws," established or necessitated we know not how.

An incalculable amount of lower life has been certainly passed through before that human organisation was attained, of which we and our generation are for the time the holders and transmitters. This is no mean heritage, and I think it should be considered as a sacred trust, for, together with man, intelligence of a sufficiently high order to produce great results appears, so far as we can infer from the varied records of the prehistoric past, to have first dawned upon the tenantry of the earth. Man has already shown his large power in the modifications he has made on the surface of the globe, and in the distribution of plants and animals. He has cleared such vast regions of forest that his work that way in North America alone, during the past half century, would be visable to an observer as far off as the moon. He has dug and drained; he has exterminated plants and animals that were mischievous to him; he has domesticated those that serve his purpose, and transplanted them to great distances from their native places. Now that this new animal man, finds himself somehow in existence, endowed with a little power and intelligence, he ought, I submit, to awake to a fuller knowledge of his relatively great position, and begin to assume a deliberate part in furthering the great work of evolution. He may infer the course it is bound to pursue, from his observation of that which it has already followed, and he might devote his modicum of power, intelligence, and kindly feeling to render its future progress less slow and painful. Man has already furthered evolution very considerably, half unconsciously, and for his own personal advantages, but he has not yet risen to the conviction that it is his religious duty to do so deliberately and systematically.

Selection and Race.

The fact of an individual being naturally gifted with high qualities, may be due either to his being an exceptionally good specimen of a poor race, or an average specimen of a high one. The difference of origin would betray itself in his descendants ; they would revert towards the typical centre of their race, deteriorating in the first case but not in the second. The two cases, though theoretically distinct, are confused in reality, owing to the frequency with which exceptional personal qualities connote the departure of the entire nature of the individual from his ancestral type, and the formation of a new strain having its own typical centre. It is hardly necessary to add that it is in this indirect way that natural selection improves a race. The two events of selection and difference of race ought, however, to be carefully distinguished in broad practical considerations, while the frequency of their concurrence is borne in mind and allowed for.

So long as the race remains radically the same, the stringent selection of the best specimens to rear and breed from, can never lead to any permanent result. The attempt to raise the standard of such a race is like the labour of Sisyphus in rolling his stone uphill ; let the effort be relaxed for a moment, and the stone will roll back. Whenever a new typical centre appears, it is as though there was a facet upon the lower surface of the stone, on which it is capable of resting without rolling back. It affords a temporary sticking-point in the forward progress of evolution.

The causes that check the unlimited improvement of highly-bred animals, so long as the race remains unchanged, are many and absolute.

In the first place there is an increasing delicacy of constitution ; the growing fineness of limb and structure end, after a few generations, in fragility. Overbred animals have little stamina ; they resemble in this respect the "weedy" colts so often reared from first-class racers. One can perhaps see in a general way why this should be so. Each individual is the outcome of a vast number of organic elements of the most various species, just as some nation might be the outcome of a vast number of castes of individuals, each caste monopolising a special pursuit. Banish a number of the humbler castes—the bakers, the bricklayers, and the smiths, and the nation would soon come to grief. This is what is done in high breeding ; certain qualities are bred for, and the rest are diminished as far as possible, but they cannot be dispensed with entirely.

The next difficulty lies in the diminished fertility of highly-bred animals. It is not improbable that its cause is of the same character as that of the delicacy of their constitution. Together with infertility is combined some degree of sexual indifference, or when passion is shown, it is not unfrequently for some specimen of a coarser type. This is certainly the case with horses and with dogs.

It will be easily understood that these difficulties, which are so formidable in the case of plants and animals, which we can mate as we please and destroy when we please, would make the maintenance of a highly-selected breed of men an impossibility.

Whenever a low race is preserved under conditions of life that exact a high level of efficiency, it must be subjected to rigorous selection. The few best specimens of that race can alone be allowed to become parents, and not many of their descendants can be allowed to live. On the other hand, if a higher race be substituted for the low one, all this terrible misery disappears. The most merciful form of what I ventured to call "eugenics" would consist in watching for the indications of superior strains or races, and in so favouring them that their progeny shall outnumber and gradually replace that of the old one. Such strains are of no infrequent occurrence. It is easy to specify families who are characterised by strong resemblances, and whose features and character are usually prepotent over those of their wives or husbands in their joint offspring, and who are at the same time as prolific as the average of their class. These strains can be conveniently studied in the families of exiles, which, for obvious reasons, are easy to trace in their various branches.

The debt that most countries owe to the race of men whom they received from one another as immigrants, whether leaving their native country of their own free will, or as exiles on political or religious grounds, has been often pointed out, and may, I think, be accounted for as follows :—The fact of a man leaving his compatriots, or so irritating them that they compel him to go, is fair evidence that either he or they, or both, feel that his character is alien to theirs. Exiles are also on the whole men of considerable force of character ; a quiet man would endure and succumb, he would not have energy to transplant himself or to become so conspicuous as to be an object of general attack. We may justly infer from this, that exiles are on the whole men of exceptional and energetic natures, and it is especially from such men as these that new strains of race are likely to proceed.

Influence of Man upon Race.

The influence of man upon the nature of his own race has already been very large, but it has not been intelligently directed, and has in many instances done great harm. Its action has been by invasions and migration of races, by war and massacre, by wholesale deportation of population, by emigration, and by many social customs which have a silent but widespread effect.

There exists a sentiment, for the most part quite unreasonable, against the gradual extinction of an inferior race. It rests on some confusion between the race and the individual, as if the destruction of a race was equivalent to the destruction of a large number of men. It is nothing of the kind when the process of extinction works silently and slowly through the earlier marriage of members of the superior race, through their greater vitality under equal stress, through their better chances of getting a livelihood, or through their prepotency in mixed marriages. That the members of an inferior class should dislike being elbowed out of the way is another matter; but it may be somewhat brutally argued that whenever two individuals struggle for a single place, one must yield, and that there will be no more unhappiness on the whole, if the inferior yield to the superior than conversely, whereas the world will be permanently enriched by the success of the superior. The conditions of happiness are, however, too complex to be disposed of by *à priori* argument; it is safest to appeal to observation. I think it could be easily shown that when the differences between the races is not so great as to divide them into obviously different classes, and where their language, education, and general interests are the same, the substitution may take place gradually without any unhappiness. Thus the movements of commerce have introduced fresh and vigorous blood into various parts of England: the new-comers have intermarried with the residents, and their characteristics have been prepotent in the descendants of the mixed marriages. I have referred in the earlier part of the book to the changes of type in the English nature that have occurred during the last few hundred years. These have been effected so silently that we only know of them by the results.

One of the most misleading of words is that of "aborigines." Its use dates from the time when the cosmogony was thought to be young and life to be of very recent appearance. Its usual meaning seems to be derived from the supposition that nations disseminated themselves like colonists from a common centre about four thousand years, say 120 generations ago, and thenceforward occupied their lands undisturbed until the very recent historic period with which the narrator deals, when some invading host drove out the "aborigines." This idyllic view of the march of events is contradicted by ancient sepulchral remains, by language, and by the habits of those modern barbarians whose history we know. There are probably hardly any spots on the earth that have not, within the last few thousand years, been tenanted by very different races ; none hardly that have not been tenanted by very different tribes having the character of at least sub-races.

The absence of a criterion to distinguish between races and sub-races, and our ethnological ignorance generally, makes it impossible to offer more than a very off-hand estimate of the average variety of races in the different countries of the world. I have, however, endeavoured to form one, which I give with much hesitation, knowing how very little it is worth. I registered the usually recognised races inhabiting each of upwards of twenty countries, and who at the same time formed at least half per cent of the population. It was, I am perfectly aware, a very rough proceeding, so rough that for the United Kingdom I ignored the prehistoric types and accepted only the three headings of British, Low Dutch, and Norman-French. Again, as regards India I registered as follows :—Forest tribes (numerous), Dravidian (three principal divisions), Early Arian, Tartar (numerous, including Afghans), Arab, and lastly European, on account of their political import-ance, notwithstanding the fewness of their numbers. Pro-ceeding in this off-hand way, and after considering the results, the broad conclusion to which I arrived was that on the average at least three different recognised races were to be found in every moderately-sized district on the earth's surface. The materials were far too scanty to enable any idea to be formed of the rate of change in the relative numbers of the constituent races in each country, and still less to estimate the secular changes of type in those races.

It may be well to take one or two examples of intermixture. Spain was occupied in the earliest historic times by at least two races, of whom we know very little ; it was afterwards colonised here and there by Phœnicians in its southern ports, and by Greeks in its eastern. In the third century B.C. it was invaded by the Carthaginians, who conquered and held a large part of it, but were afterwards supplanted by the Romans, who ruled it more or less completely for 700 years. It was invaded in the fifth century A.D. by a succession of German tribes, and was finally completely overrun by the Visigoths, who ruled it for more than 200 years. Then came the invasion of the Moors, who rapidly conquered the whole of the Peninsula up to the mountains of Asturias, where the Goths still held their own, and whence they issued from time to time and ultimately recovered the country. The present population consists of the remnants of one or more tribes of ancient Iberians, of the still more ancient Basques, and of relics of all the invaders who have just been named. There is, besides, a notable proportion of Gypsies and not a few Jews.

This is obviously a most heterogeneous mixture, but to fully appreciate the diversity of its origin the several elements should be traced farther back towards their sources. Thus, the Moors are principally descendants of Arabs, who flooded the northern provinces of Africa in successive waves of emigration eastwards, both before and after the Hegira, partly combining with the Berbers as they went, and partly displacing them from the littoral districts and driving them to the oases of the Sahara, whence they in their turn displaced the Negro population, whom they drove down to the Soudan. The Gypsies, according to Sir Henry Rawlinson,[1] came from the Indo-Scythic tribes who inhabited the mouths of the Indus, and began to migrate northward, from the fourth century onward. They settled in the Chaldean marshes, assumed independence and defied the caliph. In A.D. 831 the grandson of Haroun al-Raschid sent a large expedition against them, which, after slaughtering ten thousand, deported the whole of the remainder first to Baghdad and thence onwards to Persia. They continued unmanageable in their new home, and were finally transplanted to the Cilician frontier in Asia Minor, and established there as a military colony to guard the passes of the Taurus. In A.D. 962 the Greeks, having obtained some temporary successes, drove the Gypsies back more into the interior, whence they gradually moved towards the Hellespont under the pressure of the advancing Seljukians, during the twelfth and thirteenth centuries. They then crossed over to Europe

[1] *Proceedings of the Royal Geographical Society*, vol. i. This account of the routes of the Gypsies is by no means universally accepted, nor, indeed, was offered as a complete solution of the problem of their migration, but it will serve to show how complex that problem is.

and gradually overspread it, where they are now estimated to number more than three millions.

It must not be supposed that emigration on a large scale implies even a moderate degree of civilisation among those who emigrate, because the process has been frequently traced among the more barbarous tribes, to say nothing of the evidence largely derived from ancient burial-places.   My own impression of the races in South Africa was one of a continual state of ferment and change, of the rapid development of some clan here and of the complete or almost complete suppression of another clan there.   The well-known history of the rise of the Zulus and the destruction of their neighbours is a case in point.   In the country with which I myself was familiar the changes had been numerous and rapid in the preceding few years, and there were undoubted signs of much more important substitutions of race in bygone times.   The facts were briefly these: Damara Land was inhabited by pastoral tribes of the brown Bantu race who were in continual war with various alternations of fortune, and the several tribes had special characteristics that were readily appreciated by themselves.   On the tops of the escarped hills lived a fugitive black people speaking a vile dialect of Hottentot, and families of yellow Bushmen were found in the lowlands wherever the country was unsuited for the pastoral Damaras.   Lastly, the steadily encroaching Namaquas, a superior Hottentot race, lived on the edge of the district.   They had very much more civilisation than the Bushmen, and more than the Damaras, and they contained a large infusion of Dutch blood.

The interpretation of all this was obviously that the land had been tenanted a long time ago by Negroes, that an invasion of Bushmen drove the Negroes to the hills, and that the supremacy of these lasted so long that the Negroes lost their own language and acquired that of the Bushmen. Then an invasion of a tribe of Bantu race supplanted the Bushmen, and the Bantus, after endless struggles among themselves, were being pushed aside at the time I visited them by the incoming Namaquas, who themselves are a mixed race.   This is merely a sample of Africa ; everywhere there are evidences of changing races.

The last 300 or 400 years, say the last ten generations of mankind, have witnessed changes of population on the largest scale, by the extension of races long resident in Europe to the temperate regions of Asia, Africa, America, and Australasia.

Siberia was barely known to the Russians of nine generations ago, but since that time it has been continuously overspread by their colonists, soldiers, political exiles, and transported criminals ; already some two-thirds of its population are Sclaves.

In South Africa the settlement at the Cape of Good Hope is barely six generations old, yet during that time a curious and continuous series of changes has taken place, resulting in the substitution of an alien population for the Hottentots in the south and the Bantus in the north. One-third of it is white, consisting of Dutch, English, descendants of French Huguenot refugees, some Germans and Portuguese, and the remainder is a strange medley of Hottentot, Bantu, Malay, and Negro elements. In North Africa Egypt has become infiltrated with Greeks, Italians, Frenchmen, and Englishmen during the last two generations, and Algeria with Frenchmen.

In North America the change has been most striking, from a sparse Indian population of hunters into that of the present inhabitants of the United States and Canada ; the former of these, with its total of fifty millions inhabitants, already contains more than forty-three millions of whites, chiefly of English origin ; that is more of European blood than is to be found in any one of the five great European kingdoms of England, France, Italy, Germany, and Austria, and less than that of Russia alone. The remainder are chiefly black, the descendants of slaves imported from Africa. In the Dominion of Canada, with its much smaller population of four millions, there has been a less, but still a complete, swamping of the previous Indian element by incoming whites.

In South America, and thence upwards to Mexico inclusive, the population has been infiltrated in some parts and transformed in others, by Spanish blood and by that of the Negroes whom they introduced, so that not one half of its population can be reckoned as of pure Indian descent.

The West Indian Islands have had their population absolutely swept away since the time of the Spanish Conquest, except in a few rare instances, and African Negroes have been substituted for them.

Australia and New Zealand tell much the same tale as Canada. A native population has been almost extinguished in the former and is swamped in the latter, under the pressure of an immigrant population of Europeans, which is now twelve times as numerous as the Maories. The time during which this great change has been effected is less than that covered by three generations.

To this brief sketch of changes of population in very recent periods, I might add the wave of Arab admixture that has extended from Egypt and the northern provinces of Africa into the Soudan, and that of the yellow races of China, who have already made their industrial and social influence felt in many distant regions, and who bid fair hereafter, when certain of their peculiar religious fancies shall have fallen into decay, to become one of the most effective of the colonising nations, and who may, as I trust, extrude hereafter the coarse and lazy Negro from at least the metaliferous regions of tropical Africa.

It is clear from what has been said, that men of former generations have exercised enormous influence over the human stock of the present day, and that the average humanity of the world now and in future years is and will be very different to what it would have been if the action of our forefathers had been different. The power in man of varying the future human stock vests a great responsibility in the hands of each fresh generation, which has not yet been recognised at its just importance, nor deliberately employed. It is foolish to fold the hands and to say that nothing can be done, inasmuch as social forces and self-interests are too strong to be resisted. They need not be resisted ; they can be guided. It is one thing to check the course of a huge steam vessel by the shock of a sudden encounter when she is going at full speed in the wrong direction, and another to cause her to change her course slowly and gently by a slight turn of the helm.

Nay, a ship may be made to describe a half circle, and to end by following a course exactly opposite to the first, without attracting the notice of the passengers.

POPULATION.

Over-population and its attendant miseries may not improbably become a more serious subject of consideration than it ever yet has been, owing to improved sanatation and consequent diminution of the mortality of children, and to the filling up of the spare places of the earth which are still void and able to receive the overflow of Europe. There are no doubt conflicting possibilities which I need not stop to discuss.

The check to over-population mainly advocated by Malthus is a prudential delay in the time of marriage; but the practice of such a doctrine would assuredly be limited, and if limited it would be most prejudicial to the race, as I have pointed out in *Hereditary Genius*, but may be permitted to do so again. The doctrine would only be followed by the prudent and self-denying; it would be neglected by the impulsive and self-seeking. Those whose race we especially want to have, would leave few descendants, while those whose race we especially want to be quit of, would crowd the vacant space with their progeny, and the strain of population would thenceforward be just as pressing as before. There would have been a little relief during one or two generations, but no permanent increase of the general happiness, while the race of the nation would have deteriorated. The practical application of the doctrine of deferred marriage would therefore lead indirectly to most mischievous results, that were overlooked owing to the neglect of considerations bearing on race. While criticising the main conclusion to which Malthus came, I must take the opportunity of paying my humble tribute of admiration to his great and original work, which seems to me like the rise of a morning star before a day of free social investigation. There is nothing whatever in his book that would be in the least offensive to this generation, but he wrote in advance of his time and consequently roused virulent attacks, notably from his fellow-clergymen, whose doctrinaire notions upon the paternal dispensation of the world were rudely shocked.

The misery check, as Malthus called all those influences that are not prudential, is an ugly phrase not fully justified. It no doubt includes death through inadequate food and shelter, through pestilence from overcrowding, through war, and the like; but it also includes many causes that do not deserve so hard a name. Population decays under conditions that cannot be charged to the presence or absence of misery, in the common sense of the word. These exist when native races disappear before the presence of the incoming white man, when after making the fullest allowances for imported disease, for brandy drinking, and other assignable causes, there is always a large residuum of effect not clearly accounted for. It is certainly not wholly due to misery, but rather to listlessness, due to discouragement, and acting adversely in many ways.

One notable result of dulness and apathy is to make a person unattractive to the opposite sex and to be unattracted by them. It is antagonistic to sexual affection, and the result is a diminution of offspring. There exists strong evidence that the decay of population in some parts of South America under the irksome tyranny of the Jesuits, which crushed what little vivacity the people possessed, was due to this very cause. One cannot fairly apply the term "misery" to apathy; I should rather say that strong affections restrained from marriage by prudential considerations more truly deserved that name.

EARLY AND LATE MARRIAGES

It is important to obtain a just idea of the relative effects of early and late marriages. I attempted this in *Hereditary Genius*, but I think the following is a better estimate. We are unhappily still deficient in collected data as regards the fertility of the upper and middle classes at different ages; but the facts collected by Dr. Matthews Duncan as regards the lower orders will serve our purpose approximately, by furnishing the required *ratios*, though not the absolute values. The following are his results,[1] from returns kept at the Lying-in Hospital of St. Georges-in-the-East :—

Age of Mother at her Marriage.	Average Fertility.
15–19	9·12
20–24	7·92
25–29	6·30
30–34	4·60

The meaning of this Table will be more clearly grasped after a little modification of its contents. We may consider the fertility of each group to refer to the medium age of that group, as by writing 17 instead of 15–19, and we may slightly smooth the figures, then we have—

Age of Mother at her Marriage.	Approximate average Fertility.
17	$9·00 = 6 \times 1·5$
22	$7·50 = 5 \times 1·5$
27	$6·00 = 4 \times 1·5$
32	$4·50 = 3 \times 1·5$

which shows that the relative fertility of mothers married at the ages of 17, 22, 27, and 32 respectively is as 6, 5, 4, and 3 approximately.

The increase in population by a habit of early marriages is further augmented by the greater rapidity with which the generations follow each other. By the joint effect of these two causes, a large effect is in time produced.

Let us compute a single example. Taking a group of 100 mothers married at the age of 20, whom we will designate as A, and another group of 100 mothers married at the age of 29, whom we will call B, we shall find by interpolation that the fertility of A and B respectively would be about 8·2 and 5·4. We need not, however, regard their absolute fertility, which would differ in different classes of society, but will only consider their relative production of such female children as may live and become mothers, and we will suppose the number of such descendants in the first generation to be the same as that of the A and B mothers together

---

[1] *Fecundity, Fertility, Sterility*, etc., by Dr. Matthews Duncan. A. & C. Black : Edinburgh, 1871, p. 143.

P

—namely, 200. Then the number of such children in the A and B classes respectively, being in the proportion of 8·2 to 5·4, will be 115 and 85.

We have next to determine the average lengths of the A and B generations, which may be roughly done by basing it on the usual estimate of an average generation, irrespectively of sex, at a third of a century, or say of an average female generation at 31·5 years. We will further take 20 years as being 4·5 years earlier than the average time of marriage, and 29 years as 4·5 years later than it, so that the length of each generation of the A group will be 27 years, and that of the B group will be 36 years. All these suppositions appear to be perfectly fair and reasonable, while it may easily be shown that any other suppositions within the bounds of probability would lead to results of the same general order.

The least common multiple of 27 and 36 is 108, at the end of which term of years A will have been multiplied four times over by the factor 1·5, and B three times over by the factor 0·85. The results are given in the following Table :—

After Number of Years as below.	Number of Female Descendants who themselves become Mothers.	
	A Of 100 Mothers whose Marriages and those of their Daughters all take place at the Age of 20 Years. ——— (Ratio of Increase in each successive Generation being 1·15.)	B Of 100 Mothers whose Marriages and those of their Daughters all take place at the Age of 29 Years. ——— (Ratio of Decrease in each successive Generation being 0·85.
108	175	61
216	299	38
324	535	23

The general result is that the group B gradually disappears, and the group A more than supplants it. Hence if the races best fitted to occupy the land are encouraged to marry early, they will breed down the others in a very few generations.

MARKS FOR FAMILY MERIT

It may seem very reasonable to ask how the result proposed in the last paragraph is to be attained, and to add that the difficulty of carrying so laudable a proposal into effect lies wholly in the details, and therefore that until some working plan is suggested, the consideration of improving the human race is Utopian. But this requirement is not altogether fair, because if a persuasion of the importance of any end takes possession of men's minds, sooner or later means are found by which that end is carried into effect. Some of the objections offered at first will be discovered to be sentimental, and of no real importance—the sentiment will change and they will disappear; others that are genuine are not met, but are in some way turned or eluded; and lastly, through the ingenuity of many minds directed for a long time towards the achievement of a common purpose, many happy ideas are sure to be hit upon that would not have occurred to a single individual.

This being premised, it will suffice to faintly sketch out some sort of basis for eugenics, it being now an understanding that we are provisionally agreed, for the sake of argument, that the improvement of race is an object of first-class importance, and that the popular feeling has been educated to regard it in that light.

The final object would be to devise means for favouring individuals who bore the signs of membership of a superior race, the proximate aim would be to ascertain what those signs were, and these we will consider first.

The indications of superior breed are partly personal, partly ancestral. We need not trouble ourselves about the personal part, because full weight is already given to it in the competitive careers; energy, brain, morale, and health being recognised factors of success, while there can hardly be a better evidence of a person being adapted to his circumstances than that afforded by success. It is the ancestral part that is neglected, and which we have yet to recognise at its just value. A question that now continually arises is this: a youth is a candidate for permanent employment, his present personal qualifications are known, but how will he turn out in later years? The objections to competitive examinations are notorious, in that they give undue prominence to youths whose receptive faculties are quick, and whose intellects are precocious. They give no indication of the directions in which the health, character, and intellect of the youth will change through the development, in their due course, of ancestral tendencies that are latent in youth, but will manifest themselves in after life. Examinations deal with the present, not with the future, although it is in the future of the youth that we are especially interested. Much of the needed guidance may be derived from his family history. I cannot doubt, if two youths were of equal personal merit, of whom one belonged to a thriving and long-lived family, and the other to a decaying and short-lived family, that there could be any hesitation in saying that the chances were greater of the first-mentioned youth becoming the more valuable public servant of the two.

A thriving family may be sufficiently defined or inferred by the successive occupations of its several male members in the previous generation, and of the two grandfathers. These are patent facts attainable by almost every youth, which admit of being verified in his neighbourhood and attested in a satisfactory manner.

A healthy and long-lived family may be defined by the patent facts of ages at death, and number and ages of living relatives, within the degrees mentioned above, all of which can be verified and attested. A knowledge of the existence of longevity in the family would testify to the stamina of the candidate, and be an important addition to the knowledge of his present health in forecasting the probability of his performing a large measure of experienced work.

Owing to absence of data and the want of inquiry of the family antecedents of those who fail and of those who succeed in life, we are much more ignorant than we ought to be of their relative importance. In connection with this, I may mention some curious results published by Mr. F. M. Holland[1] of Boston, U.S., as to the antecedent family history of persons who were reputed to be more moral than the average, and of those who were the reverse. He has

[1] *Index Newspaper*, Boston, U.S. July 27, 1882.

been good enough to reply to questions that I sent to him concerning his criterion of morality, and other points connected with the statistics, in a way that seems satisfactory, and he has very obligingly furnished me with additional MS. materials. One of his conclusions was that morality is more often found among members of large families than among those of small ones. It is reasonable to expect this would be the case owing to the internal discipline among members of large families, and to the wholesome sustaining and restraining effects of family pride and family criticism. Members of small families are apt to be selfish, and when the smallness of the family is due to the deaths of many of its members at early ages, it is some evidence either of weakness of the family constitution, or of deficiency of common sense or of affection on the part of the parents in not taking better care of them. Mr. Holland quotes in his letter to me a piece of advice by Franklin to a young man in search of a wife, "to take one out of a bunch of sisters," and a popular saying that kittens brought up with others make the best pets, because they have learned to play without scratching. Sir William Gull[1] has remarked that those candidates for the Indian Civil Service who are members of large families are on the whole the strongest.

Far be it from me to say that any scheme of marks for family merit would not require a great deal of preparatory consideration. Careful statistical inquiries have yet to be made into the family antecedents of public servants of mature age in connection with their place in examination lists at the earlier age when they first gained their appointments. This would be necessary in order to learn the amount of marks that should be assigned to various degrees of family merit. I foresee no peculiar difficulty in conducting such an inquiry ; indeed, now that competitive examinations have been in general use for many years, the time seems ripe for it, but of course its conduct would require much confidential inquiry and a great deal of trouble in verifying returns. Still, it admits of being done, and if the results, derived from different sources, should confirm one another, they could be depended on.

[1] *Blue Book* C—1446, 1876. On the Selection and Training of Candidates for the Indian Civil Service.

Let us now suppose that a way was seen for carrying some such idea as this into practice, and that family merit, however defined, was allowed to count, for however little, in competitive examinations. The effect would be very great: it would show that ancestral qualities are of present current value; it would give an impetus to collecting family histories; it would open the eyes of every family and or society at large to the importance of marriage alliance with a good stock; it would introduce the subject of race into a permanent topic of consideration, which (on the supposition of its *bonâ fide* importance that has been assumed for the sake of argument) experience would show to be amply justified. Any act that first gives a guinea stamp to the sterling guinea's worth of natural nobility might set a great social avalanche in motion.

ENDOWMENTS.

Endowments and bequests have been freely and largely made for various social purposes, and as a matter of history they have frequently been made to portion girls in marriage. It so happens that the very day that I am writing this, I notice an account in the foreign newspapers (September 19, 1882) of an Italian who has bequeathed a sum to the corporation of London to found small portions for three poor girls to be selected by lot. And again, a few weeks ago I read also in the French papers of a trial, in reference to the money adjudged to the "Rosière" of a certain village. Many cases in which individuals and states have portioned girls may be found in Malthus. It is therefore far from improbable that if the merits of good race became widely recognised and its indications were rendered more surely intelligible than they now are, that local endowments, and perhaps adoptions, might be made in favour of those or both sexes who showed evidences of high race and of belonging to prolific and thriving families. One cannot forecast their form, though we may reckon with some assurance that in one way or another they would be made, and that the better races would be given a better chance or marrying early.

A curious relic of the custom which was universal three or

four centuries ago, of entrusting education to celibate priests, forbade Fellows of Colleges to marry, under the penalty of losing their fellowships. It is as though the winning horses at races were rendered ineligible to become sires, which I need hardly say is the exact reverse of the practice. Races were established and endowed by "Queen's plates" and otherwise at vast expense, for the purpose of discovering the swiftest horses, who are thenceforward exempted from labour and reserved for the sole purpose of propagating their species. The horses who do not win races, or who are not otherwise specially selected for their natural gifts, are prevented from becoming sires. Similarly, the mares who win races as fillies, are not allowed to waste their strength in being ridden or driven, but are tended under sanatory conditions for the sole purpose of bearing offspring. It is better economy, in the long-run, to use the best mares as breeders than as workers, the loss through their withdrawal from active service being more than recouped in the next generation through what is gained by their progeny.

The college statutes to which I referred were very recently relaxed at Oxford, and have been just reformed at Cambridge. I am told that numerous marriages have ensued in consequence, or are ensuing. In *Hereditary Genius* I showed that scholastic success runs strongly in families; therefore, in all seriousness, I have no doubt, that the number of Englishmen naturally endowed with high scholastic faculties, will be sensibly increased in future generations by the repeal of these ancient statutes.

The English race has yet to be explored and their now unknown wealth of hereditary gifts recorded, that those who possess such a patrimony should know of it. The natural impulses of mankind would then be sufficient to ensure that such wealth should no more continue to be neglected than the existence of any other possession suddenly made known to a man. Aristocracies seldom make alliances out of their order, except to gain wealth. Is it less to be expected that those who become aware that they are endowed with the power of transmitting valuable hereditary gifts should abstain from squandering their future children's patrimony by marrying persons of lower natural stamp? The social consideration that would attach itself to high races would, it may be hoped, partly neutralise a social cause that is now very adverse to the early marriages of the most gifted, namely, the cost of living in cultured and refined society. A young man with a career before him commonly feels it would be an act of folly to hamper himself by too early a marriage. The doors of society that are freely open to a bachelor are closed to a married couple with small means, unless they bear patent recommendations such as the public recognition of a natural nobility would give. The attitude of mind that I should expect to predominate among those who had undeniable claims to rank as members of an exceptionally gifted race, would be akin to that of the modern possessors of ancestral property or hereditary rank. Such persons feel it a point of honour not to alienate the old place or make misalliances, and they are respected for their honest family pride. So a man of good race would shrink from spoiling it by a lower marriage, and every one would sympathise with his sentiments.

CONCLUSION.

It remains to sketch in outline the principal conclusions to which we seem to be driven by the results of the various inquiries contained in this volume, and by what we know on allied topics from the works of others.

We cannot but recognise the vast variety of natural faculty, useful and harmful, in members of the same race, and much more in the human family at large, all of which tend to be transmitted by inheritance. Neither can we fail to observe that the faculties of men generally, are unequal to the requirements of a high and growing civilisation. This is principally owing to their entire ancestry having lived·up to recent times under very uncivilised conditions, and to the somewhat capricious distribution in late times of inherited wealth, which affords various degrees of immunity from the usual selective agencies.

In solution of the question whether a continual improvement in education might not compensate for a stationary or even retrograde condition of natural gifts, I made inquiry into the life history of twins, which resulted in proving the vastly preponderating effects of nature over nurture.

The fact that the very foundation and outcome of the human mind is dependent on race, and that the qualities of races vary, and therefore that humanity taken as a whole is not fixed but variable, compels us to reconsider what may be the true place and function of man in the order of the world. I have examined this question freely from many points of view, because whatever may be the vehemence with which particular opinions are insisted upon, its solution is unquestionably doubtful. There is a wide and growing conviction among truth-seeking, earnest, humble-minded, and thoughtful men, both in this country and abroad, that our cosmic relations are by no means so clear and simple as they are popularly supposed to be, while the worthy and intelligent teachers of various creeds, who have strong persuasions on the character of those relations, do not concur in their several views.

The results of the inquiries I have made into certain alleged forms of our relations with the unseen world do not, so far as they go, confirm the common doctrines. One, for example, on the objective efficacy of prayer [1] was decidedly negative. It showed that while contradicting the commonly expressed doctrine, it concurred with the almost universal practical opinion of the present day. Another inquiry into visions showed that, however ill explained they may still be, they belong for the most part, if not altogether, to an order of phenomena which no one dreams in other cases of calling supernatural. Many investigations concur in showing the vast multiplicity of mental operations that are in simultaneous action, of which only a minute part falls within the ken of consciousness, and suggest that much of what passes for supernatural is due to one portion of our mind being contemplated by another portion of it, as if it had been that of another person. The term " individuality " is in fact a most misleading word.

I do not for a moment wish to imply that the few inquiries published in this volume exhaust the list of those that might be made, for I distinctly hold the contrary, but

[1] Not reprinted in this edition.

I refer to them in corroboration of the previous assertion that our relations with the unseen world are different to those we are commonly taught to believe.

In our doubt as to the character of our mysterious relations with the unseen ocean of actual and potential life by which we are surrounded, the generally accepted fact of the solidarity of the universe—that is, of the intimate connections between distant parts that bind it together as a whole—justifies us, I think, in looking upon ourselves as members of a vast system which in one of its aspects resembles a cosmic republic.

On the one hand, we know that evolution has proceeded during an enormous time on this earth, under, so far as we can gather, a system of rigorous causation, with no economy of time or of instruments, and with no show of special ruth for those who may in pure ignorance have violated the conditions of life.

On the other hand, while recognising the awful mystery of conscious existence and the inscrutable background of evolution, we find that as the foremost outcome of many and long birth-throes, intelligent and kindly man finds himself in being. He knows how petty he is, but he also perceives that he stands here on this particular earth, at this particular time, as the heir of untold ages and in the van of circumstance. He ought therefore, I think, to be less diffident than he is usually instructed to be, and to rise to the conception that he has a considerable function to perform in the order of events, and that his exertions are needed. It seems to me that he should look upon himself more as a freeman, with power of shaping the course of future humanity, and that he should look upon himself less as the subject of a despotic government, in which case it would be his chief merit to depend wholly upon what had been regulated for him, and to render abject obedience.

The question then arises as to the way in which man can assist in the order of events. I reply, by furthering the course of evolution. He may use his intelligence to discover and expedite the changes that are necessary to adapt circumstance to race and race to circumstance, and his kindly sympathy will urge him to effect them mercifully.

When we begin to inquire, with some misgiving perhaps, as to the evidence that man has present power to influence the quality of future humanity, we soon discover that his past influence in that direction has been very large indeed. It has been exerted hitherto for other ends than that which is now contemplated, such as for conquest or emigration, also through social conditions whose effects upon race were imperfectly foreseen. There can be no doubt that the hitherto unused means of his influence are also numerous and great. I have not cared to go much into detail concerning these, but restricted myself to a few broad considerations, as by showing how largely the balance of population becomes affected by the earlier marriages of some of its classes, and by pointing out the great influence that endowments have had in checking the marriage of monks and scholars, and therefore the yet larger influence they might be expected to have if they were directed not to thwart but to harmonise with natural inclination, by promoting early marriages in the classes to be favoured. I also showed that a powerful influence might flow from a public recognition in early life of the true value of the probability of future performance, as based on the past performance of the ancestors of the child. It is an element of forecast, in addition to that of present personal merit, which has yet to be appraised and recognised. Its recognition would attract assistance in various ways, impossible now to specify, to the young families of those who were most likely to stock the world with healthy, moral, intelligent, and fair-natured citizens. The stream of charity is not unlimited, and it is requisite for the speedier evolution of a more perfect humanity that it should be so distributed as to favour the best-adapted races. I have not spoken of the repression of the rest, believing that it would ensue indirectly as a matter of course; but I may add that few would deserve better of their country than those who determine to live celibate lives, through a reasonable conviction that their issue would probably be less fitted than the generality to play their part as citizens.

It would be easy to add to the number of possible agencies by which the evolution of a higher humanity might be furthered, but it is premature to do so until the importance of attending to the improvement of our race shall have been so well established in the popular mind that a discussion of them would be likely to receive serious consideration.

It is hardly necessary to insist on the certainty that our present imperfect knowledge of the limitations and conditions of hereditary transmission will be steadily added to; but I would call attention again to the serious want of adequate materials for study in the form of life-histories. It is fortunately the case that many of the rising medical practitioners of the foremost rank are become strongly impressed with the necessity of possessing them, not only for the better knowledge of the theory of disease, but for the personal advantage of their patients, whom they now have to treat less appropriately than they otherwise would, through ignorance of their hereditary tendencies and of their illnesses in past years, the medical details of which are rarely remembered by the patient, even if he ever knew them. With the help of so powerful a personal motive for keeping life-histories, and of so influential a body as the medical profession to advocate its being done,[1] and to show how to do it, there is considerable hope that the want of materials to which I have alluded will gradually be supplied.

To sum up in a few words. The chief result of these Inquiries has been to elicit the religious significance of the doctrine of evolution. It suggests an alteration in our mental attitude, and imposes a new moral duty. The new mental attitude is one of a greater sense of moral freedom, responsibility, and opportunity; the new duty which is supposed to be exercised concurrently with, and not in opposition to the old ones upon which the social fabric depends, is an endeavour to further evolution, especially that of the human race.

[1] See an address on the Collective Investigation of Disease, by Sir William Gull, *British Medical Journal*, January 27, 1883, p. 143; also the following address by Sir James Paget, p. 144.

APPENDIX

A.—COMPOSITE PORTRAITURE.

THE object and methods of Composite Portraiture will be best explained by the following extracts from memoirs describing its successive stages, published in 1878, 1879, and 1881 respectively :—

I. COMPOSITE PORTRAITS, MADE BY COMBINING THOSE OF MANY DIFFERENT PERSONS INTO A SINGLE RESULTANT FIGURE.

[Extract from Memoir read before the Anthropological Institute, in 1878.]

I submit to the Anthropological Institute my first results in carrying out a process that I suggested last August [1877] in my presidential address to the Anthropological Subsection of the British Association at Plymouth, in the following words :—

"Having obtained drawings or photographs of several persons alike in most respects, but differing in minor details, what sure method is there of extracting the typical characteristics from them? I may mention a plan which had occurred both to Mr. Herbert Spencer and myself, the principle of which is to superimpose optically the various drawings, and to accept the aggregate result. Mr. Spencer suggested to me in conversation that the drawings reduced to the same scale might be traced on separate pieces of transparent paper and secured one upon another, and then held between the eye and the light. I have attempted this with some success. My own idea was to throw faint images of the several portraits, in succession, upon the same sensitised photographic plate. I may add that it is perfectly easy to superimpose optically two portraits by means of a stereoscope, and that a person who is used to handle instruments will find a common double eyeglass fitted with stereoscopic lenses to be almost as effectual and far handier than the boxes sold in shops."

Mr. Spencer, as he informed me, had actually devised an instrument, many years ago, for tracing mechanically, longitudinal, transverse, and horizontal sections of heads on transparent paper, intending to superimpose them, and to obtain an average result by transmitted light.

Since my address was published, I have caused trials to be made, and have found, as a matter of fact, that the photographic process of which I there spoke enables us to obtain with mechanical precision a generalised picture; one that represents no man in particular, but portrays an imaginary figure possessing the average features of any given group of men. These ideal faces have a surprising air of reality. Nobody who glanced at one of them for the first time would doubt its being the likeness of a living person, yet, as I have said, it is no such thing ; it is the portrait of a type and not of an individual.

I begin by collecting photographs of the persons with whom I propose to deal. They must be similar in attitude and size, but no exactness is necessary in either of these respects. Then, by a simple contrivance, I make two pinholes in each of them, to enable me to hang them up one in front of the other, like a

pack of cards, upon the same pair of pins, in such a way that the eyes of all the portraits shall be as nearly as possible superimposed ; in which case the remainder of the features will also be superimposed nearly enough. These pinholes correspond to what are technically known to printers as " register marks." They are easily made : A slip of brass or card has an aperture cut out of its middle, and threads are stretched from opposite sides, making a cross.[1] Two small holes are drilled in the plate, one on either side of the aperture. The slip of brass is laid on the portrait with the aperture over its face. It is turned about until one of the cross threads cuts the pupils of both the eyes, and it is further adjusted until the other thread divides the interval between the pupils in two equal parts. Then it is held firmly, and a prick is made through each of the holes.

The portraits being thus arranged, a photographic camera is directed upon them. Suppose there are eight portraits in the pack, and that under existing circumstances it would require an

---

[1] I am indebted for the woodcuts to the Editor of *Nature*, in which journal this memoir first appeared.

exposure of eighty seconds to give an exact photographic copy of any one of them. The general principle of proceeding is this, subject in practice to some variations of detail, depending on the different brightness of the several portraits. We throw the image of each of the eight portraits in turn upon the same part of the sensitised plate for ten seconds. Thus, portrait No. 1 is in the front of the pack ; we take the cap off the object glass of the camera for ten seconds, and afterwards replace it. We then remove No. 1 from the pins, and No. 2 appears in the front ; we take off the cap a second time for ten seconds, and again replace it. Next we remove No. 2, and No. 3 appears in the front, which we treat as its predecessors, and so we go on to the last of the pack. The sensitised plate will now have had its total exposure of eighty seconds ; it is then developed, and the print taken from it is the generalised picture of which I

speak. It is a composite of eight component portraits. Those of its outlines are sharpest and darkest that are common to the largest number of the components ; the purely individual peculiarities leave little or no visible trace. The latter being necessarily disposed equally on both sides of the average, the outline of the composite is the average of all the components. It is a band and not a fine line, because the outlines of the components are seldom exactly superimposed. The band will be darkest in its middle whenever the component portraits have the same general type of features, and its breadth, or amount of blur, will measure the tendency of the components to deviate from the common type. This is so for the very same reason that the shot-marks on a target are more thickly disposed near the bull's-eye than away from it, and in a greater degree as the marksmen are more skilful. All that has been said of the outlines is equally true as regards the shadows ; the result being that the composite represents an averaged figure, whose lineaments have been softly drawn. The eyes come out with appropriate distinctness, owing to the mechanical conditions under which the components are hung.

A composite portrait represents the picture that would rise before the mind's eye of a man who had the gift of pictorial imagination in an exalted degree. But the imaginative power even of the highest artists is far from precise, and is so apt to be biassed by special cases that may have struck their fancies, that no two artists agree in any of their typical forms. The merit of the photographic composite is its mechanical precision, being subject to no errors beyond those incidental to all photographic productions.

I submit several composites made for me by Mr. H. Reynolds. The first set of portraits are those of criminals convicted of murder, manslaughter, or robbery accompanied with violence. It will be observed that the features of the composites are much better looking than those of the components. The special villainous irregularities in the latter have disappeared, and the common humanity that underlies them has prevailed. They represent, not the criminal, but the man who is liable to fall into crime. All composites are better looking than their components, because the averaged portrait of many persons is free from the irregularities that variously blemish the looks of each of them.

I selected these for my first trials because I happened to possess a large collection of photographs of criminals, through the kindness of Sir Edmund Du Cane, the Director-General of Prisons, for the purpose of investigating criminal types. They were peculiarly adapted to my present purpose, being all made of about the same size, and taken in much the same attitudes. It was while endeavouring to elicit the principal criminal types by methods of optical superimposition of the portraits, such as I had frequently employed with maps and meteorological traces,[1] that the idea of composite figures first occurred to me.

The other set of composites are made from pairs of components. They are selected to show the extraordinary facility of combining almost any two faces whose proportions are in any way similar.

It will, I am sure, surprise most persons to see how well defined these composites are. When we deal with faces of the same type, the points of similarity far outnumber those of dissimilarity, and there is a much greater resemblance between faces generally than we who turn our attention to individual differences are apt to appreciate. A traveller on his first arrival among people of a race very different to his own thinks them closely alike, and a Hindu has much difficulty in distinguishing one Englishman from another.

The fairness with which photographic composites represent

[1] *Conference at the Loan Exhibition of Scientific Instruments*, 1878. Chapman and Hall. Physical Geography Section, p. 312, *On Means of Combining Various Data in Maps and Diagrams*, by Francis Galton, F.R.S.

their components is shown by six of the specimens. I wished to learn whether the order in which the components were photographed made any material difference in the result, so I had three of the portraits arranged successively in each of their six possible combinations. It will be observed that four at least of the six composites are closely alike. I should say that in each of this set (which was made by the wet process) the last of the three components was always allowed a longer exposure

The accompanying woodcut is as fair a representation of one of the composites as is practicable in ordinary printing. It was photographically transferred to the wood, and the engraver has used his best endeavour to translate the shades into line engraving. This composite is made out of only three components, and its threefold origin is to be traced in the ears, and in the buttons to the vest. To the best of my judgment, the original photograph is a very exact average of its components; not one feature in it appears identical with that of any one of them, but it contains a resemblance to all, and is not more like to one of them than to another. However, the judgment of the wood engraver is different. His rendering of the composite has made it exactly like one of its components, which it must be borne in mind he had never seen. It is just as though an artist drawing a child had produced a portrait closely resembling its deceased father, having overlooked an equally strong likeness to its deceased mother, which was apparent to its relatives. This is to me a most striking proof that the composite is a true combination.

than the second, and the second than the first, but it is found better to allow an equal time to all of them.

The stereoscope, as I stated last August in my address at Plymouth, affords a very easy method of optically superimposing two portraits, and I have much pleasure in quoting the following letter, pointing out this fact as well as some other conclusions to which I also had arrived. The letter was kindly forwarded to me by Mr. Darwin; it is dated last November, and was written to him by Mr. A. L. Austin, from New Zealand, thus affording another of the many curious instances of two persons being independently engaged in the same novel inquiry at nearly the same time, and coming to similar results :—

"INVERCARGILL, NEW ZEALAND,
"*November 6th*, 1877.

" To CHARLES DARWIN, Esq.

" SIR,—Although a perfect stranger to you, and living on the reverse side of the globe, I have taken the liberty of writing to you on a small discovery I have made in binocular vision in the stereoscope. I find by taking two ordinary carte-de-visite photos of two different persons' faces, the portraits being about the same sizes, and looking about the same direction, and placing them in a stereoscope, the faces blend into one in a most remarkable manner, producing in the case of some ladies' portraits, in every instance, a *decided improvement* in beauty. The pictures were not taken in a binocular camera, and therefore do not stand out well, but by moving one or both until the eyes coincide in the stereoscope the pictures blend perfectly. If taken in a binocular camera for the purpose, each person being taken on one half of the negative, I am sure the results would be still more striking. Perhaps something might be made of this in regard to the expression of emotions in man and the lower animals, &c. I have not time or opportunities to make experiments, but it seems to me something might be made of this by photographing the faces of different animals, different races of mankind, &c. I think a stereoscopic view of one of the ape tribe and some low-caste human face would make a very curious mixture ; also in the matter of crossing of animals and the resulting offspring. It seems to me something also might result in photos of husband and wife and children, &c. In any case, the results are curious, if it leads to nothing else. Should this come to anything you will no doubt acknowledge myself as suggesting the experiment, and perhaps send me some of the results. If not likely to come to anything, a reply would much oblige me.

" Yours very truly,
" A. L. AUSTIN, C.E., F.R.A.S."

Dr. Carpenter informs me that the late Mr. Appold, the mechanician, used to combine two portraits of himself under the stereoscope. The one had been taken with an assumed stern expression, the other with a smile, and this combination produced a curious and effective blending of the two.

Convenient as the stereoscope is, owing to its accessibility, for determining whether any two portraits are suitable in size and attitude to form a good composite, it is nevertheless a makeshift and imperfect way of attaining the required result. It cannot of itself combine two images ; it can only place them so that the office of attempting to combine them may be undertaken by the brain. Now the two separate impressions received by the brain through the stereoscope do not seem to me to be relatively constant in their vividness, but sometimes the image seen by the left eye prevails over that seen by the right, and *vice versâ.* All the other instruments I am about to describe accomplish that which the stereoscope fails to do ; they create true optical combinations. As regards other points in Mr. Austin's letter, I cannot think that the use of a binocular camera for taking the two portraits intended to be combined into one by the stereoscope would be of importance. All that is wanted is that the portraits should be nearly of the same size. In every other respect I cordially agree with Mr. Austin.

The best instrument I have as yet[4] contrived and used for optical superimposition is a "double-image prism" of Iceland spar (see Fig., p. 228), formerly procured for me by the late Mr. Tisley, optician, Brompton Road. They have a clear aperture of a square, half an inch in the side, and when held at right angles to the line of sight will separate the ordinary and extraordinary images to the amount of two inches, when the object viewed is held at seventeen inches from the eye. This is quite sufficient for working with carte-de-visite portraits. One image is quite achromatic, the other shows a little colour. The divergence may be varied and adjusted by inclining the prism to the line of sight. By its means the ordinary image of one component is thrown upon the extraordinary image of the other, and the composite may be viewed by the naked eye, or through a lens of long focus, or through an opera-glass (a telescope is not so good) fitted with a sufficiently long draw-tube to see an object at that short distance with distinctness. Portraits of somewhat different sizes may be combined by placing the larger one farther from the eye, and a long face may be fitted to a short one by inclining and foreshortening the former. The slight fault of focus thereby occasioned produces little or no sensible ill effect on the appearance of the composite.

The front, or the profile, faces of two living persons sitting side by side or one behind the other, can be easily superimposed by a double-image prism. Two such prisms set one behind the other can be made to give four images of equal brightness, occupying the four corners of a rhombus whose acute angles are 45°. Three prisms will give eight images, but this is practically not a good combination ; the images fail in distinctness, and are too near together for use. Again, each lens of a stereoscope of long focus can have one or a pair of these prisms attached to it, and four or eight images may be thus combined.

Another instrument I have made consists of a piece of glass inclined at a very acute angle to the line of sight, and of a

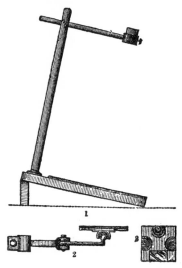

Fig. 1 shows the simple apparatus which carries the prism and on which the photograph is mounted. The former is set in a round box which can be rotated in the ring at the end of the arm and can be clamped when adjusted. The arm can be rotated and can also be pulled out or in if desired, and clamped. The floor of the instrument is overlaid with cork covered with black cloth, on which the components can easily be fixed by drawing-pins. When using it, one portrait is pinned down and the other is moved near to it, overlapping its margin if necessary, until the eye looking through the prism sees the required combination; then the second portrait is pinned down also. It may now receive its register-marks from needles fixed in a hinged arm, and this is a more generally applicable method than the plan with cross threads, already described, as any desired feature—the nose, the ear, or the hand, may thus be selected for composite purposes. Let A, B, C, . . . Y, Z, be the components. A is pinned down, and B, C, . . . Y, Z, are successfully combined with A, and registered. Then before removing Z, take away A and substitute any other of the already registered portraits, say B, by combining it with Z; lastly, remove Z and substitute A by combining it with B, and register it. Fig. 2 shows one of three similarly jointed arms, which clamp on to the vertical rod. Two of these carry a light frame covered with cork and cloth, and the other carries Fig. 3, which is a frame having lenses of different powers set into it, and on which, or on the third frame, a small mirror inclined at 45° may be laid. When a portrait requires foreshortening it can be pinned on one of these frames and be inclined to the line of sight; when it is smaller than its fellow it can be brought nearer to the eye and an appropriate lens interposed; when a right-sided profile has to be combined with a left-handed one, it must be pinned on one of the frames and viewed by reflection from the mirror in the other. The apparatus I have drawn is roughly made, and being chiefly of wood is rather clumsy, but it acts well.

mirror beyond it, also inclined, but in the opposite direction to the line of sight. Two rays of light will therefore reach the eye from each point of the glass ; the one has been reflected from its surface, and the other has been first reflected from the mirror, and then transmitted through the glass. The glass used should be extremely thin, to avoid the blur due to double reflections ; it may be a selected piece from those made to cover microscopic specimens. The principle of the instrument may be yet further developed by interposing additional pieces of glass, successively less inclined to the line of sight, and each reflecting a different portrait.

I have tried many other plans ; indeed the possible methods of optically superimposing two or more images are very numerous. Thus I have used a sextant (with its telescope attached) ; also strips of mirrors placed at different angles, their several reflections being simultaneously viewed through a telescope. I have also used a divided lens, like two stereoscopic lenses brought close together, in front of the object glass of a telescope.

II. GENERIC IMAGES.

[Extract from Proceedings Royal Institution, 25th April 1879.]

Our general impressions are founded upon blended memories, and these latter will be the chief topic of the present discourse. An analogy will be pointed out between these and the blended portraits first described by myself a year ago under the name of " Composite Portraits," and specimens of the latter will be exhibited.

The physiological basis of memory is simple enough in its broad outlines. Whenever any group of brain elements has been excited by a sense impression, it becomes, so to speak, tender, and liable to be easily thrown again into a similar state of excitement. If the new cause of excitement differs from the original one, a memory is the result. Whenever a single cause throws different groups of brain elements simultaneously into excitement, the result must be a blended memory.

We are familiar with the fact that faint memories are very apt to become confused. Thus some picture of mountain and lake in a country which we have never visited, often recalls a vague sense of identity with much we have seen elsewhere. Our recollections cannot be disentangled, though general resemblances are recognised. It is also a fact that the memories of persons who have great powers of visualising, that is, of seeing well-defined images in the mind's eye, are no less capable of being blended together. Artists are, as a class, possessed of the visualising power in a high degree, and they are at the same time pre-eminently distinguished by their gifts of generalisation. They are of all men the most capable of producing forms that are not copies of any individual, but represent the characteristic features of classes.

There is then, no doubt, from whatever side the subject of memory is approached, whether from the material or from the mental, and, in the latter case, whether we examine the experiences of those in whom the visualising faculty is faint or in whom it is strong, that the brain has the capacity of blending memories together. Neither can there be any doubt that general impressions are faint and perhaps faulty editions of blended memories. They are subject to errors of their own, and they inherit all those to which the memories are themselves liable.

Specimens of blended portraits will now be exhibited ; these might, with more propriety, be named, according to the happy phrase of Professor Huxley, "generic" portraits. The word generic presupposes a genus, that is to say, a collection of individuals who have much in common, and among whom medium characteristics are very much more frequent than extreme ones. The same idea is sometimes expressed by the word "typical," which was much used by Quetelet, who was the first to give it a rigorous interpretation, and whose idea of a type lies at the basis of his statistical views. No statistician dreams of combining objects into the same generic group that do not cluster towards a common centre ; no more should we attempt to compose generic portraits out of heterogeneous elements, for if we do so the result is monstrous and meaningless.

It might be expected that when many different portraits are fused into a single one, the result would be a mere smudge. Such, however, is by no means the case, under the conditions just laid down, of a great prevalence of the mediocre characteristics over the extreme ones. There are then so many traits in common, to combine and to reinforce one another, that they prevail to the exclusion of the rest. All that is common remains, all that is individual tends to disappear.

The first of the composites exhibited on this occasion is made by conveying the images of three separate portraits by means of three separate magic-lanterns upon the same screen. The stands on which the lanterns are mounted have been arranged to allow of nice adjustment. The composite about to be shown is one that strains the powers of the process somewhat too severely, the portraits combined being those of two brothers and their sister, who have not even been photographed in precisely the same attitudes. Nevertheless, the result is seen to be the production of a face, neither male nor female, but more regular and handsome than any of the component portraits, and in which the common family traits are clearly marked. Ghosts of portions of male and female attire, due to the peculiarities of the separate portraits, are seen about and around the composite, but they are not sufficiently vivid to distract the attention. If the number of combined portraits had been large, these ghostly accessories would have become too faint to be visible.

The next step is to compare this portrait of two brothers and their sister which has been composed by optical means before the eyes of the audience, and concerning the truthfulness of which there can be no doubt, with a photographic composite of the same group. The latter is now placed in a fourth magic-lantern with a brighter light behind it, and its image is thrown on the screen by the side of the composite produced by direct optical superposition. It will be observed that the two processes lead to almost exactly the same result, and therefore the fairness of the photographic process may be taken for granted. However, two other comparisons will be made for the sake of verification, namely, between the optical and photographic composites of two children, and again between those of two Roman contadini.

The composite portraits that will next be exhibited are made by the photographic process, and it will now be understood that they are truly composite, notwithstanding their definition and apparent individuality. Attention is, however, first directed to a convenient instrument not more than 18 inches in length, which is, in fact, a photographic camera with six converging lenses and an attached screen, on which six pictures can be adjusted and brilliantly illuminated by artificial light. The effect of their optical combination can thus be easily studied ; any errors of adjustment can be rectified, and the composite may be photographed at once.

It must not be supposed that any one of the components fails to leave its due trace in the photographic composite, much less in the optical one. In order to allay misgivings on the subject, a small apparatus is laid on the table together with some of the results obtained by it. It is a cardboard frame, with a spring shutter closing an aperture of the size of a wafer, that springs open on the pressure of a finger, and shuts again as suddenly when the pressure is withdrawn. A chronograph is held in the other hand, whose index begins to travel the moment the finger presses a spring, and stops instantly on lifting the finger. The two instruments are worked simultaneously ; the chronograph checking the time allowed for each exposure and summing all the times. It appears from several trials that the effect of 1000 brief exposures is practically identical with that of a single exposure of 1000 times the duration of any one of them. Therefore each of a thousand components leaves its due photographic trace on the composite, though it is far too faint to be visible unless reinforced by many similar traces.

The composites now to be exhibited are made from coins or medals, and in most instances the aim has been to obtain the best likeness attainable of historical personages, by combining various portraits of them taken at different periods of their lives, and so to elicit the traits that are common to each series. A few of the individual portraits are placed in the same slide with each composite to give a better idea of the character of these blended representatives. Those that are shown are (1) Alexander the Great, from six components; (2) Antiochus, King of Syria, from six; (3) Demetrius Poliorcetes, from six; (4) Cleopatra, from five. Here the composite is as usual better looking than any of the components, none of which, however, give any indication of her reputed beauty; in fact, her features are not only plain, but to an ordinary English taste are simply hideous. (5) Nero, from eleven; (6) A combination of five different Greek female faces; and (7) A singularly beautiful combination of the faces of six different Roman ladies, forming a charming ideal profile.

My cordial acknowledgment is due to Mr. R. Stuart Poole, the learned curator of the coins and gems in the British Museum, for his kind selection of the most suitable medals, and for procuring casts of them for me for the present purpose. These casts were, with one exception, all photographed to a uniform size of four-tenths of an inch between the pupils of the eyes and the division between the lips, which experience shows to be the most convenient size on the whole to work with, regard being paid to many considerations not worth while to specify in detail. When it was necessary the photograph was reversed. These photographs were made by Mr. H. Reynolds; I then adjusted and prepared them for taking the photographic composite.

The next series to be exhibited consists of composites taken from the portraits of criminals convicted of murder, manslaughter, or crimes accompanied by violence. There is much interest in the fact that two types of features are found much more frequently among these than among the population at large. In one, the features are broad and massive, like those of Henry VIII., but with a much smaller brain. The other, of which five composites are exhibited, each deduced from a number of different individuals, varying four to nine, is a face that is weak and certainly not a common English face. Three of these composites, though taken from entirely different sets of individuals, are as alike as brothers, and it is found on optically combining any three out of the five composites, that is on combining almost any considerable number of the individuals, the result is closely the same. The combination of the three composites just alluded to will now be effected by means of the three converging magic-lanterns, and the result may be accepted as generic in respect of this particular type of criminals.

The process of composite portraiture is one of pictorial statistics. It is a familiar fact that the average height of even a dozen men of the same race, taken at hazard, varies so little, that for ordinary statistical purposes it may be considered constant. The same may be said of the measurement of every separate feature and limb, and of every tint, whether of skin, hair, or eyes. Consequently a pictorial combination of any one of these separate traits would lead to results no less constant than the statistical averages. In a portrait, there is another factor to be considered besides the measurement of the separate traits, namely, their relative position ; but this, too, in a sufficiently large group, would necessarily have a statistical constancy. As a matter of observation, the resemblance between persons of the same " genus " (in the sense of "generic," as already explained) is sufficiently great to admit of making good pictorial composites out of even small groups, as has been abundantly shown.

Composite pictures, are, however, much more than averages ; they are rather the equivalents of those large statistical tables whose totals, divided by the number of cases, and entered in the bottom line, are the averages. They are real generalisations, because they include the whole of the material under consideration. The blur of their outlines, which is never great in truly generic composites, except in unimportant details, measures the tendency of individuals to deviate from the central type. My argument is, that the generic images that arise before the mind's eye, and the general impressions which are faint and faulty editions of them, are the analogues of these composite pictures which we have the advantage of examining at leisure, and whose peculiarities and character we can investigate, and from which we may draw conclusions that shall throw much light on the nature of certain mental processes which are too mobile and evanescent to be directly dealt with.

III. Composite Portraiture.

[Read before the Photographic Society, 24th June, 1881.]

propose to draw attention to-night to the results of recent experiments and considerable improvements in a process of which I published the principles three years ago, and which I have subsequently exhibited more than once.

I have shown that, if we have the portraits of two or more different persons, taken in the same aspect and under the same conditions of light and shade, and that if we put them into different optical lanterns converging on the same screen and carefully adjust them—first, so as to bring them to the same scale, and, secondly, so as to superpose them as accurately as the conditions admit—then the different faces will blend surprisingly well into a single countenance. If they are not very dissimilar, the blended result will always have a curious air of individuality, and will be unexpectedly well defined; it will exactly resemble none of its components, but it will have a sort of family likeness to all of them, and it will be an ideal and an averaged portrait. I have also shown that the image on the screen might be photographed then and there, or that the same result may be much more easily obtained by a method of successive photography, and I have exhibited many specimens made on this principle. Photo-lithographs of some of these will be found in the *Proceedings of the Royal Institution*, as illustrations of a lecture I gave there " On Generic Images " in 1879.

The method I now use is much better than those previously described; it leads to more accurate results, and is easier to manage. I will exhibit and explain the apparatus as it stands, and will indicate some improvements as I go on. The apparatus is here. I use it by gaslight, and employ rapid dry plates, which, however, under the conditions of a particularly small aperture and the character of the light, require sixty seconds of total exposure. The apparatus is 4 feet long and $6\frac{1}{2}$ inches broad; it lies with its side along the edge of the table at which I sit, and it is sloped towards me, so that, by bending my neck slightly, I can bring my eye to an eye-hole, where I watch the effect of the adjustments which my hands are free to make. The entire management of the whole of these is within an easy arm's length, and I complete the process without shifting my seat.

The apparatus consists of three parts, A, B, and C. A is rigidly fixed; it contains the dark slide and the contrivances by which the position of the image can be viewed; the eye-hole, *e*, already mentioned, being part of A. B is a travelling carriage that holds the lens, and is connected by bellows-work with A. In my apparatus it is pushed out and in, and clamped where desired, but it ought to be moved altogether by pinion and rack-work.[1] The lens I use is a I B Dallmeyer. Its focal length is appropriate to the size of the instrument, and I find great convenience in a lens of wide aperture when making the adjustments, as I then require plenty of light; but, as to the photography, the smaller the aperture the better. The hole in my stop is only two-tenths of an inch in diameter, and I believe one-tenth would be more suitable.

[1] I have since had a more substantial instrument made with these and similar improvements.

C is a travelling carriage that supports the portraits in turn, from which the composite has to be made. I work directly from the original negatives with transmitted light; but prints

DIAGRAM SHOWING THE ESSENTIAL PARTS

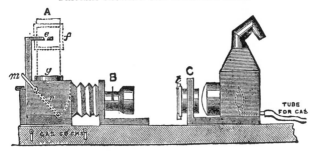

Side View.

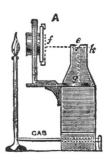

End View.

A the body of the camera, which is fixed.

B Lens on a carriage, which can be moved to and fro.

C Frame for the transparency, on a carriage that also supports the lantern; the whole can be moved to and fro.

r The reflector inside the camera.

m The arm outside the camera attached to the axis of the reflector; by moving it, the reflector can be moved up or down.

g A ground-glass screen on the roof, which receives the image when the reflector is turned down, as in the diagram.

e The eye-hole through which the image is viewed on g; a thin piece of glass immediately below e, reflects the illuminated fiducial lines in the transparency at f, and gives them the appearance of lying upon g,— the distances f e and g e being made equal, the angle f e g being made a right angle, and the plane of the thin piece of glass being made to bisect f e g.

f Framework, adjustable, holding the transparency with the fiducial lines on it.

t Framework, adjustable, holding the transparency of the portrait.

can be used with light falling on their face. For convenience of description I will confine myself to the first instance only, and will therefore speak of C as the carriage that supports the frame that holds the negative transparencies. C can be pushed along the board and be clamped anywhere, and it has a rack and pinion adjustment ; but it should have been made movable by rack and pinion along the whole length of the board. The frame for the transparencies has the same movements of adjustment as those in the stage of a microscope. It rotates round a hollow axis, through which a beam of light is thrown, and independent movements in the plane, at right angles to the axis, can be given to it in two directions, at right angles to one another, by turning two separate screws. The beam of light is furnished by three gas-burners, and it passes through a condenser. The gas is supplied through a flexible tube that does not interfere with the movements of C, and it is governed by a stop-cock in front of the operator.

The apparatus, so far as it has been described with any detail, and ignoring what was said about an eye-hole, is little else than a modified copying-camera, by which an image of the transparency could be thrown on the ordinary focusing-screen, and be altered in scale and position until it was adjusted to fiducial lines drawn on the screen. It is conceivable that this should be done, and that the screen should be replaced by the dark slide, and a brief exposure given to the plate ; then, that a fresh transparency should be inserted, a fresh focusing adjustment made, and a second exposure given, and so on. This, I say, is conceivable, but it would be very inconvenient. The adjusting screws would be out of reach ; the head of the operator would be in an awkward position ; and though these two difficulties might be overcome in some degree, a serious risk of an occasional shift of the plate during the frequent replacement of the dark slide would remain. I avoid all this by making my adjustments while the plate continues in position with its front open. I do so through the help of a reflector temporarily interposed between it and the lens. I do not use the ordinary focusing-screen at all in making my adjustments, but one that is flush, or nearly so, with the roof of the camera. When the reflector is interposed, the image is wholly cut off from the sensitised plate, and is thrown upwards against this focusing-screen, g. When the reflector is withdrawn, the image falls on the plate. It is upon this focusing-screen in the roof that I see the fiducial lines by which I make all the adjustments. Nothing can be more convenient than the position of this focusing-screen for working purposes. I look down on the image as I do upon a book resting on a sloping desk, and all the parts of the apparatus are within an easy arm's length.

My reflector in my present instrument is, I am a little ashamed to confess, nothing better than a piece of looking-glass fixed to an axle within the camera, near its top left-hand edge. One end of the axle protrudes, and has a short arm; when I push the arm back, the mirror is raised; when I push it forward it drops down. I used a swing-glass because the swing action is very true, and as my apparatus was merely a provisional working model made of soft wood, I did not like to use sliding arrangements which might not have acted truly, or I should certainly have employed a slide with a rectangular glass prism, on account of the perfect reflection it affords. And let me say, that a prism of 2 inches square in the side is quite large enough for adjustment purposes, for it is only the face of the portrait that is wanted to be seen. I chose my looking-glass carefully, and selected a piece that was plane and parallel. It has not too high a polish, and therefore does not give troublesome double reflections. In fact, it answers very respectably, especially when we consider that perfection of definition is thrown away on composites. I thought of a mirror silvered on the front of the glass, but this would soon tarnish in the gaslight, so I did not try it. For safety against the admission of light unintentionally, I have a cap to the focusing-screen in the roof, and a slide in the fixed body of the instrument immediately behind the reflector and before the dark slide. Neither of these would be wanted if the reflector was replaced by a prism, set into one end of a sliding block that had a large horizontal hole at the other end, and a sufficient length of solid wood between the two to block out the passage of light both upwards and downwards whenever the block is passing through the half-way position.

As regards the fiducial lines, they might be drawn on the glass screen; but black lines are not, I find, the best. It is far easier to work with illuminated lines; and it is important to be able to control their brightness. I produce these lines by means of a vertical transparency, set in an adjustable frame, connected with A, and having a gas-light behind it. Below the eye-hole e, through which I view the glass-screen g, is a thin piece of glass set at an angle of $45°$, which reflects the fiducial lines and gives them the appearance of lying on the screen, the frame being so adjusted that the distance from the thin piece of glass to the transparency and to the glass-screen g is the same. I thus obtain beautiful fiducial lines, which I can vary from extreme faintness to extreme brilliancy, by turning the gas lower or higher, according to the brightness of the image of the portrait, which itself depends on the density of the transparency that I am engaged upon. This arrangement seems as good as can be. It affords a gauge of the density of the negative, and enables me to regulate the burners behind it, until the image of the portrait on g is adjusted to a standard degree of brightness.

For convenience in enlarging or reducing, I take care that the intersection of the vertical fiducial line with that which passes through the pupils of the eyes shall correspond to the optical axis of the camera. Then, as I enlarge or reduce, that point in the image remains fixed. The uppermost horizontal fiducial line continues to intersect the pupils, and the vertical one continues to divide the face symmetrically. The mouth has alone to be watched. When the mouth is adjusted to the lower fiducial line, the scale is exact. It is a great help having to attend to no more than one varying element. The only inconvenience is that the image does not lie in the best position on the plate when the point between the eyes occupies its centre. This is easily remedied by using a larger back with a suitable inner frame. I have a more elaborate contrivance in my apparatus to produce the same result, which I need not stop to explain.

For success and speed in making composites, the apparatus should be solidly made, chiefly of metal, and all the adjustments ought to work smoothly and accurately. Good composites cannot be made without very careful adjustment in scale and position. An off-hand way of working produces nothing but failures.

I will first exhibit a very simple but instructive composite effect. I drew on a square card a circle of about $2\frac{1}{2}$ inches in diameter, and two cross lines through its centre, cutting one another at right angles. Round each of the four points, 90° apart, where the cross cuts the circle, I drew small circles of the size of wafers and gummed upon each a disc of different tint. Finally I made a single black dot half-way between two of the arms of the cross. I then made a composite of the four positions of the card, as it was placed successively with each of its sides downwards. The result is a photograph having a sharply-defined cross surrounded by four discs of precisely uniform tint, and between each pair of arms of the cross there is a very faint dot. This photograph shows many things. The fact of its being a composite is shown by the four faint dots. The equality of the successive periods of exposure is shown by the equal tint of the four dots. The accuracy of adjustment is shown by the sharpness of the cross being as great in the composite as in the original card. We see the smallness of the effect produced by any trait, such as the dot, when it appears in the same place in only one of the components : if this effect be so small in a series of only four components, it would certainly be imperceptible in a much larger series. Thirdly, the uniformity of resulting tint in the composite wafer is quite irrespective of the order of exposure. Let us call the four component wafers A, B, C, D, respectively, and the four composite wafers 1, 2, 3, 4; then we see, by the diagram, that the order of exposure has differed in each case.

Composite.		Successive places of the Components.						
1	2	A	B	D	A	C	D	B	C
4	3	D	C	C	B	B	A	A	D

In 1 it has been A, D, C, B,
„ 2   „   B, A, D, C,
„ 3   „   C, B, A, D,
„ 4   „   D, C, B, A,

yet the result is identical. Therefore the order of exposure has no effect on the result.

I will next show a series consisting of two portraits considerably unlike to one another, and yet not so very discordant as to refuse to conform, and of two intermediate composites. In making one of the composites I gave two-thirds of the total time of exposure to the first portrait, and one-third to the second portrait. In making the other composite, I did the converse. It will be seen how good is the result in both cases, and how the likeness of the longest exposed portrait always predominates.

The next is a series of four composites. The first consists of 57 hospital patients suffering under one or other of the many forms of consumption. I may say that, with the aid of Dr. Mahomed, I am endeavouring to utilise this process to elicit the physiognomy of disease. The composite I now show is what I call a hotch-pot composite; its use is to form a standard whence deviations towards any particular sub-type may be conveniently gauged. It will be observed that the face is strongly marked, and that it is quite idealised. I claim for composite portraiture, that it affords a method of obtaining *pictorial averages*, which effects simultaneously for every point in a picture what a method of numerical averages would do for each point in the picture separately. It gives, in short, the average tint of every unit of area in the picture, measured from the fiducial lines as co-ordinates. Now every statistician knows, by experience, that numerical averages usually begin to agree pretty fairly when we deal with even twenty or thirty cases. Therefore we should expect to find that any groups of twenty or thirty men of the same class would yield composites bearing a considerable likeness to one another. In proof that this is the case, I exhibit three other composites: the one is made from the first 28 portraits of the 57, the second from the last 27, and the third is made from 36 portraits taken indiscriminately out of the 57. It will be observed that all the four composites are closely alike.

I will now show a few typical portraits I selected out of 82 male portraits of a different series of consumptive male patients ; they were those that had more or less of a particular wan look, that I wished to elicit. The selected cases were about 18 in number, and from these I took 12, rejecting about six as having some marked peculiarity that did not conform well with the remaining 12. The result is a very striking face, thoroughly ideal and artistic, and singularly beautiful. It is, indeed, most notable how beautiful all composites are. Individual peculiarities are all irregularities, and the composite is always regular.

I show a composite of 15 female faces, also of consumptive patients, that gives somewhat the same aspect of the disease ; also two others of only 6 in each, that have in consequence less of an ideal look, but which are still typical. I have here several other typical faces in my collection of composites ; they are all serviceable as illustrations of this memoir, but, medically speaking, they are only provisional results.

I am indebted to Lieutenant Leonard Darwin, R.E., for an interesting series of negatives of officers and privates of the Royal Engineers. Here is a composite of 12 officers ; here is one of 30 privates. I then thought it better to select from the latter the men that came from the southern counties, and to again make a further selection of 11 from these, on the principle already explained. Here is the result. It is very interesting to note the stamp of culture and refinement on the composite officer, and the honest and vigorous but more homely features of the privates. The combination of these two, officers and privates together, gives a very effective physiognomy.

Let it be borne in mind that existing cartes-de-visite are almost certain to be useless. Among dozens of them it is hard to find three that fulfil the conditions of similarity of aspect and of shade. The negatives have to be made on purpose. I use a repeating back and a quarter plate, and get two good-sized heads on each plate, and of a scale that never gives less than four-tenths of an inch between the pupils of the eyes and the mouth. It is only the head that can be used, as more distant parts, even the ears, become blurred hopelessly.

It will be asked, of what use can all this be to ordinary photographers, even granting that it may be of scientific value in ethnological research, in inquiries into the physiognomy of disease, and for other special purposes? I think it can be turned to most interesting account in the production of family likenesses. The most unartistic productions of amateur photography do quite as well for making composites as those of the best professional workers, because their blemishes vanish in the blended result. All that amateurs have to do is to take negatives of the various members of their families in precisely the same aspect (I recommend either perfect full-face or perfect profile), and under precisely the same conditions of light and shade, and to send them to a firm provided with proper instrumental appliances to make composites from them. The result is sure to be artistic in expression and flatteringly handsome, and would be very interesting to the members of the family. Young and old, and persons of both sexes can be combined into one ideal face. I can well imagine a fashion setting in to have these pictures.

Professional skill might be exercised very effectively in retouching composites. It would be easy to obliterate the ghosts of stray features that are always present when the composite is made from only a few portraits, and it would not be difficult to tone down any irregularity in the features themselves, due to some obtrusive peculiarity in one of the components. A higher order of artistic skill might be well bestowed upon the composites that have been made out of a large number of components. Here the irregularities disappear, the features are perfectly regular and idealised, but the result is dim. It is like a pencil drawing, where many attempts have been made to obtain the desired effect ; such a drawing is smudged and ineffective ; but the artist, under its guidance, draws his final work with clear bold touches, and then he rubs out the smudge. On precisely the same principle the faint but beautifully idealised features of these composites are, I believe, capable of forming the basis of a very high order of artistic work.

B.—THE RELATIVE SUPPLIES FROM TOWN AND COUNTRY FAMILIES TO THE POPULATION OF FUTURE GENERATIONS.

[Read before the Statistical Society in 1873.]

It is well known that the population of towns decays, and has to be recruited by immigrants from the country, but I am not aware that any statistical investigation has yet been attempted of the rate of its decay. The more energetic members of our race, whose breed is the most valuable to our nation, are attracted from the country to our towns. If residence in towns seriously interferes with the maintenance of their stock, we should expect the breed of Englishmen to steadily deteriorate, so far as that particular influence is concerned.

I am well aware that the only perfectly trustworthy way of conducting the inquiry is by statistics derived from numerous life-histories, but I find it very difficult to procure these data. I therefore have had recourse to an indirect method, based on a selection from the returns made at the census of 1871, which appears calculated to give a fair approximation to the truth. My object is to find the number of adult male representatives in this generation, of 1000 adult males in the previous one, of rural and urban populations respectively. The principle on which I have proceeded is this :—

I find (A) the number of children of equal numbers of urban and of rural mothers. The census schedules contain returns of the names and ages of the members of each " family," by which word we are to understand those members who are alive and resident in the same house with their parents. When the mothers are young, the children are necessarily very young, and nearly always (in at least those classes who are unable to send their children to boarding schools) live at home. If, therefore, we limit our inquiries to the census "families" of young mothers, the results may be accepted as practically identical with those we should have obtained if we had direct means of ascertaining the number of their living children. The limits of age of the mothers which I adopted in my selection were, 24 and 40 years. Had I to begin the work afresh, I should prefer the period from 20 to 35, but I have reason to feel pretty well contented with my present data. I correct the results thus far obtained on the following grounds :—(B) the relative mortality of the two classes between childhood and maturity ; (C) the relative mortality of the rural and urban mothers during childbearing ages ; (D) their relative celibacy ; and (E) the span of a rural and urban generation. It will be shown that B is important, and C noteworthy, but that D and E may be disregarded.

In deciding on the districts to be investigated, it was important to choose well-marked specimens of urban and rural populations. In the former, a town was wanted where there were various industries, and where the population was not increasing. A town where only one industry was pursued would not be a fair sample, because the particular industry might be suspected of having a special influence, and a town that was increasing would have attracted numerous immigrants from the country, who are undistinguishable as such in the census returns. Guided by these considerations, I selected Coventry, where silk weaving, watch-making, and other industries are carried on, and whose population had scarcely varied during the decade preceding the census of 1871.[1] It is an open town, in which the crowded alleys of larger places are not frequent. Its urban peculiarities are therefore minimised, and its statistical returns would give a picture somewhat too favourable of the average condition of life in towns. For specimens of rural districts, I chose small agricultural parishes in Warwickshire.

By the courteous permission of Dr. Farr, I was enabled to procure extracts from the census returns concerning 1000 "families" of factory hands at Coventry, in which the age of the mother was neither less than 24 nor more than 40 years, and concerning another 1000 families of agricultural labourers in rural parishes of Warwickshire, under the same limitations as to the age of the mother. When these returns were classified (see Table I., p. 246), I found the figures to run in such regular sequence as to make it certain that the cases were sufficiently numerous to give trustworthy results. It appeared that :—

(A) The 1000 families of factory hands comprised 2681 children, and the 1000 of agricultural labourers comprised 2911 ; hence, the children in the urban "families," the mothers being between the ages of 24 and 40, are on the whole about 8 per cent. less numerous than the rural. I see no reason why these numbers should not be accepted as relatively correct for families, in the ordinary sense of that word, and for mothers of all ages. An inspection of the table does indeed show that if the selection had begun at an earlier age than 24, there would have been an increased proportion of sterile and of small families among the factory hands, but not sufficient to introduce any substantial modification of the above results. It is, however, important to recollect that the small error, whatever its amount may be, is a concession in favour of the towns.

(B) I next make an allowance for the mortality between childhood and maturity, which will diminish the above figures in different proportions, because the conditions of town life are more fatal to children than those of the country. No life tables exist for Coventry and Warwickshire ; I am therefore obliged to use statistics for similarly conditioned localities, to determine the amount of the allowance that should be made. The life tables of Manchester[2] will afford the data for towns, and those of the "Healthy Districts"[3] will suffice for the country. By applying these, we could calculate the number of the children of ages specified in the census returns who would attain maturity. I regret extremely that when I had the copies taken, I did not give instructions to have the ages of all the children inserted ; but I did not, and it is too late now to remedy the omission. I

1 It has greatly changed since this was written.
2 "Seventh Annual Report of Registrar-General."
3 Healthy Districts Life Table, by Dr. Farr. *Phil. Trans. Royal Society,* 1859.

am therefore obliged to make a very rough, but not unfair, estimate. The average age of the children was about 3 years, and 25 years may be taken as representing the age of maturity. Now it will be found that 74 per cent. of children in Manchester, of the age of 3, reach the age of 25, while 86 per cent. of children do so in the "Healthy Districts." Therefore, if my rough method be accepted as approximately fair, the number of adults who will be derived from the children of the 1000 factory families should be reckoned at $\left(2681 \times \dfrac{74}{100}\right) = 1986$, and those from the 1000 agricultural at $\left(2911 \times \dfrac{86}{100}\right) = 2503$.

(C) The comparison we seek is between the total families produced by an equal number of urban and rural women who had survived the age of 24. Many of these women will not marry at all; I postpone that consideration to the next paragraph. Many of the rest will die before they reach the age of 40, and more of them will die in the town than in the country. It appears from data furnished by the above-mentioned tables, that if 100 women of the age of 24 had annually been added to a population, the number of those so added, living between the ages of 24 and 40 (an interval of seventeen years) would be 1539 under the conditions of life in Manchester, and 1585 under those of the healthy districts. Therefore the small factors to be applied respectively to the two cases, on account of this correction, are $\dfrac{1539}{17 \times 100}$ and $\dfrac{1585}{17 \times 100}$.

(D) I have no trustworthy data for the relative prevalence of celibacy in town and country. All that I have learned from the census returns is, that when searching them for the 1000 families, 131 bachelors were noted between the ages of 24 and 40, among the factory hands, and 144 among the agricultural labourers. If these figures be accepted as correct guides to the amount of celibacy among the women, it would follow that I must be considered to have discussed the cases of 1131 factory, and 1144 agricultural women, when dealing with those of 1000 mothers in either class. Consequently that the respective corrections to be applied, are given by the factors $\dfrac{1000}{1131}$ and $\dfrac{1000}{1141}$, or $\dfrac{88\cdot4}{1000}$, and $\dfrac{87\cdot6}{1000}$. This difference of less than 1 per cent. is hardly worth applying, moreover I do not like to apply it, because it seems to me erroneous and to act in the wrong direction, inasmuch as unmarried women can obtain employment more readily in the town than in the country, and celibacy is therefore more likely to be common in the former than in the latter.

(E) The possible difference in the length of an urban and

rural generation must not be forgotten. We, however, have reason to believe that the correction on this ground will be insignificant, because the length of a generation is found to be constant under very different circumstances of race, and therefore we should expect it to be equally constant in the same race under different conditions ; such as it is, it would probably tell against the towns.

Let us now sum up the results. The corrections are not to be applied for (D) and (E), so we have only to regard (A) × (B) × (C), that this—

$$\frac{2681 \times \dfrac{74}{100} \times \dfrac{1539}{1700}}{2911 \times \dfrac{86}{100} \times \dfrac{1585}{1700}} = \frac{1796}{2334} = \frac{77}{100}.$$

In other words, the rate of supply in towns to the next adult generation is only 77 per cent., or, say, three-quarters of that in the country. This decay, if it continued constant, would lead to the result that the representatives of the townsmen would be less than half as numerous as those of the country folk after one century, and only about one fifth as numerous after two centuries, the proportions being $\frac{45}{100}$ and $\frac{21}{100}$ respectively.

TABLE I.—*Census Returns of* 1000 *Families of Factory Hands in Coventry, according to the Age of the Mother and*

Age of Mother.	NUMBER OF CHILDREN IN FAMILY.									
	0.		1.		2.		3.		4.	
	Factory.	Agricultural.	Factory.	Agricultural.	Factory.	Agricultural.	Factory.	Agricultural.	Factory.	Agricultural.
24 to 25 . . .	28	17	40	31	24	32	12	10	2	...
26 ,, 27 . . .	19	18	36	24	36	28	23	26	8	8
28 ,, 29 . . .	18	17	32	16	20[1]	33	36	23	14	23
30 ,, 31 . . .	13	4	23	18	24	21	28[1]	31	18	22
32 ,, 33 . . .	18	11	16	14	19	13	22[1]	27	23	26
34 ,, 35 . . .	14	15	11	6	17	16	28	18	31	34
36 ,, 37 . . .	12	17	4	11	10	13	22	14	16	20
38 ,, 39 . . .	8	6	9	15	14	17	16	21	22	23
40	8	7	3	10	8	9	13	14	8	10
Total within outline .	96	67	158	109	116	111	171	149	...	...
Total between outlines	42	45	16	36	56	71	29	35	142	166
Total beyond outline .	...	...	...	...	...	...	...	...	...	...
Total . . .	138	112	174	145	172	182	200	184	142	166

[1] These three cases are anomalous, the Factory being less than the Agricultural. In that neither of these can be correct ; certainly not the first of them.

Note.—It will be observed to the left of the outline, that is, in the upper and left hand predominate, while the agricultural are the most numerous between the outlines, that is are from four to five in number. The two are equally numerous to the right of the outlines,

and 1000 *Families of Agricultural Labourers in Warwickshire, grouped*
the Number of Children in the Family.

NUMBER OF CHILDREN IN FAMILY.										
5.		6.		7.		8.		9.		Age of Mother.
Factory.	Agricultural.	Factory.	Agricultural.	Factory.	Agricultural.	Factory.	Agricultural.	Factory.	Agricultural.	
I	1	...	...	...	...	...	...	...	...	24 to 25
...	...	...	...	...	...	...	...	...	...	26 ,, 27
6	6	4	1	2	...	...	...	...	...	28 ,, 29
12	15	2	5	...	2	...	1	...	...	30 ,, 31
21	25	9	5	...	1	...	2	...	...	32 ,, 33
14	18	12	9	5	3	...	1	...	...	34 ,, 35
15	25	12	10	4	5	5	2	...	...	36 ,, 37
14	22	10	15	6	7	...	2	I	...	38 ,, 39
7	11	3	9	7	7	2	1	...	...	40
...	...	...	...	...	...	...	...	...	...	Total within outline.
90	123	...	...	...	...	...	...	...	...	Total between outlines.
...	...	52	54	24	25	7	9	I	...	Total beyond outline.
90	123	52	54	24	25	7	9	I	...	Total.

the instance of 20-33, the anomaly is double, because the sequence of the figures shows

of the table, where the mothers are young and the children few, the factory families
especially in the middle of the table, where the mothers are less young, and the families
that is, to the right of the table, where the families are large.

TABLE II.

	Number of Families.		Number of Children.	
	Factory.	Agricultural.	Factory.	Agricultural.
Within outline .	541	436	903	778
Between outlines .	375	476	1233	1562
Beyond ,, .	84	88	545	571
Total . .	1000	1000	2681	2911

C.—AN APPARATUS FOR TESTING THE DELICACY WITH WHICH WEIGHTS CAN BE DISCRIMINATED BY HANDLING THEM.

[*Read at the Anthropological Institute, Nov.* 14, 1882.]

I submit a simple apparatus that I have designed to measure the delicacy of the sensitivity of different persons, as shown by their skill in discriminating weights, identical in size, form, and colour, but different in specific gravity. Its interest lies in the accordance of the successive test values with the successive graduations of a true scale of sensitivity, in the ease with which the tests are applied, and the fact that the same principle can be made use of in testing the delicacy of smell and taste.

I use test-weights that mount in a series of "just perceptible differences" to an imaginary person of extreme delicacy of perception, their values being calculated according to Weber's law. The lowest weight is heavy enough to give a decided sense of weight to the hand when handling it, and the heaviest weight can be handled without any sense of fatigue. They therefore conform with close approximation to a geometric series; thus—

$$WR^0, \quad WR^1, \quad WR^2, \quad WR^3, \text{ etc.,}$$

and they bear as register-marks the values of the successive indices, 0, 1, 2, 3, etc. It follows that if a person can just distinguish between any particular pair of weights, he can also just distinguish between any other pair of weights whose register-marks differ by the same amount. Example: suppose A can just distinguish between the weights bearing the register-marks 2 and 4, then it follows from the construction of the apparatus that he can just distinguish between those bearing the register-marks 1 and 3, or 3 and 5, or 4 and 6, etc.; the difference being 2 in each case.

There can be but one interpretation of the phrase that the dulness of muscular sense in any person, B, is twice as great as in that of another person, A. It is that B is only capable of perceiving one grade of difference where A can perceive two. We may, of course, state the same fact inversely, and say that the delicacy of muscular sense is in that case twice as great in A as in B. Similarly in all other cases of the kind. Conversely, if having known nothing previously about either A or B, we discover on trial that A can just distinguish between two weights such as those bearing the register-marks 5 and 7, and that B can just distinguish between another pair, say, bearing the register-marks 2 and 6; then since the difference between the marks in the latter case is twice as great as in the former, we know that the dulness of the muscular sense of B is exactly twice that of A. Their relative dulness, or if we prefer to speak in inverse terms, and say their relative sensitivity, is determined quite independently of the particular pair of weights used in testing them.

It will be noted that the conversion of results obtained by the use of one series of test-weights into what would have been given by another series, is a piece of simple arithmetic, the fact ultimately obtained by any apparatus of this kind being the "just distinguishable" fraction of real weight. In my own apparatus the unit of weight is 2 per cent.; that is, the register-mark 1 means 2 per cent.; but I introduce weights in the earlier part of the scale that deal with half units; that is, with differences of 1 per cent. In another apparatus the unit of weight might be 3 per cent., then three grades of mine would be equal to two of the other, and mine would be converted to that scale by multiplying them by $\frac{2}{3}$. Thus the results obtained by different apparatus are strictly comparable.

A sufficient number of test-weights must be used, or trials made, to eliminate the influence of chance. It might perhaps be thought that by using a series of only five weights, and requiring them to be sorted into their proper order by the sense of touch alone, the chance of accidental success would be too small to be worth consideration. It might be said that there are 5 × 4 × 3 × 2, or 120 different ways in which five weights can be arranged, and as only one is right, it must be 120 to 1 against a lucky hit. But this is many fold too high an estimate, because the 119 possible mistakes are by no means equally probable. When a person is tested, an approximate value for his grade of sensitivity is rapidly found, and the inquiry becomes narrowed to finding out whether he can surely pass a particular level. At this stage of the inquiry there is little fear of a gross mistake. He is little likely to make a mistake of double the amount in question, and it is almost certain that he will not make a mistake of treble the amount. In other words, he would never be likely to put one of the test-weights more than one step out of its proper place. If he had three weights to arrange in their consecutive order, 1, 2, 3, there are $3 \times 2 = 6$ ways of arranging them ; of these, he would be liable to the errors of 1, 3, 2, and of 2, 1, 3, but he would hardly be liable to such gross errors as 2, 3, 1, or 3, 2, 1, or 3, 1, 2. Therefore of the six permutations in which three weights may be arranged three have to be dismissed from consideration, leaving three cases only to be dealt with, of which two are wrong and one is right. For the same reason there are only four reasonable chances of error in arranging four weights, and only six in arranging five weights, instead of the 119 that were originally supposed. These are—

$$12354 \quad 13245 \quad 13254$$
$$21345 \quad 21354 \quad 21435$$

But exception might be taken to two even of these, namely, those that appear in the third column, where 5 is found in juxtaposition with 2 in the first case, and 4 with 1 in the second. So great a difference between two adjacent weights would be almost sure to attract the notice of the person who was being tested, and make him dissatisfied with the arrangement. Considering all this, together with the convenience of carriage and manipulation, I prefer to use trays, each containing only three weights, the trials being made three or four times in succession. In each trial there are three possibilities and only one success, therefore in three trials the probabilities against uniform success are as 27 to 1, and in four trials at 81 to 1.

Values of the Weights.—After preparatory trials, I adopted 1000 grains as the value of W and 1020 as that of R, but I am now inclined to think that 1010 would have been better. I made the weights by filling blank cartridges with shot, wool, and wads, so as to distribute the weight equally, and I closed the cartridges with a wad, turning the edges over it with the instrument well known to sportsmen. I wrote the corresponding value of the index of R on the wad by which each of them was closed, to serve as a register number. Thus the cartridge whose weight was WR^4 was marked 4. The values were so selected that there should be as few varieties as possible. There are thirty weights in all, but only ten varieties, whose Register Numbers are respectively 0, 1, 2, 3, $3\frac{1}{2}$, $4\frac{1}{2}$, 5, 6, 7, 9, 12. The reason of this limitation of varieties was to enable the weights to be interchanged whenever there became reason to suspect that the eye had begun to recognise the appearance of any one of them, and that the judgment might be influenced by that recognition, and cease to be wholly guided by the sense of weight.

We are so accustomed to deal with concurrent impressions that it is exceedingly difficult, even with the best intention of good faith, to ignore the influence of any corroborative impression that may be present. It is therefore right to take precautions against this possible cause of inaccuracy. The most perfect way would be to drop the weights, each in a little bag or sheath of light material, so that the operatee could not see the weights, while the ratio between the weights would not be sensibly changed by the additional weight of the bags. I keep little bags for this purpose, inside the box that holds the weights.

Arrangement of the Weights.—The weights are placed in sets of threes, each set in a separate shallow tray, and the trays lie in two rows in a box. Each tray bears the register-marks of each of the weights it contains. It is also marked boldly with a Roman numeral showing the difference between the register-marks of the adjacent weights. This difference indicates the grade of sensitivity that the weights in the tray are designed to test. Thus the tray containing the weights WR^0, WR^3, WR^6 is marked as in Fig. 1, and that which contains WR^2, WR^7, WR^{12} is marked as in Fig. 2.

III.
o, 3, 6.

Fig. 1.

V.
2, 7, 12.

Fig. 2.

The following is the arrangement of the trays in the box. The triplets they contain suffice for ordinary purposes.

Just perceptible Ratio.	Grade of Sensitivity.	Sequences of Weights.	Just perceptible Ratio	Grade of Sensitivity.	Sequences of Weights.
1·020	I.	1, 2, 3	1·030	I.$\frac{1}{2}$	2, 3$\frac{1}{2}$, 5
1·040	II.	3, 5, 7	1·050	II.$\frac{1}{2}$	2, 4$\frac{1}{2}$, 7
1·061	III.	0, 3, 6	1·071	III.$\frac{1}{2}$	0, 3$\frac{1}{2}$, 7
1·082	IV.	1, 5, 9	1·082	IV.$\frac{1}{2}$	0, 4$\frac{1}{2}$, 9
1·104	V.	2, 5, 7	1·127	VI.	0, 6, 12

But it will be observed that sequences of $\frac{1}{2}$ can also be obtained, and again, that it is easy to select doublets of weights for coarser tests, up to a maximum difference of XII., which may be useful in cases of morbidly diminished sensitivity.

Manipulation.—A tray is taken out, the three weights that it contains are shuffled by the operator, who then passes them on to the experimenter. The latter sits at ease with his hand in an unconstrained position, and lifts the weights in turn between his finger and thumb, the finger pressing against the top, the thumb against the bottom of the cartridge. Guided by the touch alone, he arranges them in the tray in what he conceives to be their proper sequence ; he then returns the tray to the operator, who notes the result, the operator then reshuffles the weights and repeats the trial. It is necessary to begin with coarse preparatory tests, to accustom the operatee to the character of the work. After a minute or two the operator may begin to record results, and the testing may go for several minutes, until the hand begins to tire, the judgment to be confused, and blunders to arise. Practice does not seem to increase the delicacy of perception after the first few trials, so much as might be expected.

D.—WHISTLES FOR TESTING THE UPPER LIMITS OF AUDIBLE SOUND IN DIFFERENT INDIVIDUALS.

The base of the inner tube of the whistle is the foremost end of a plug, that admits of being advanced or withdrawn by screwing it out or in ; thus the depth of the inner tube of the whistle can be varied at pleasure. The more nearly the plug is screwed home, the less is the depth of the whistle and the more shrill does its note become, until a point is reached at which, although the air that proceeds from it vibrates as violently as before, as shown by its effect on a sensitive flame, the note ceases to be audible.

The number of vibrations per second in the note of a whistle or other "closed pipe" depends on its depth. The theory of acoustics shows that the length of each complete vibration is four times that of the depth of the closed pipe, and since experience proves that all sound, whatever may be its pitch, is propagated at the same rate, which under ordinary conditions of temperature and barometric pressure may be taken at 1120 feet, or 13,440 inches per second,—it follows that the number of vibrations in the note of a whistle may be found by dividing 13,440 by four times the depth, measured in inches, of the inner tube of the whistle. This rule, however, supposes the vibrations of the air in the tube to be strictly longitudinal, and ceases to apply when the depth of the tube is less than about one and a half times its diameter. When the tube is reduced to a shallow pan, a note may still be produced by it, but that note has reference rather to the diameter of the whistle than to its depth, being sometimes apparently unaltered by a further decrease of depth. The necessity of preserving a fair proportion between the diameter and the depth of a whistle is the reason why these instruments, having necessarily little depth, require to be made with very small bores.

The depth of the inner tube of the whistle at any moment is shown by the graduations on the outside of the instrument. The lower portion of the instrument as formerly made for me by the late Mr. Tisley, optician, Brompton Road,[1] is a cap that surrounds the body of the whistle, and is itself fixed to the screw that forms the plug. One complete turn of the cap increases or diminishes the depth of the whistle, by an amount equal to the interval between two adjacent threads of the screw. For mechanical convenience, a screw is used whose pitch is 25 to the inch ; therefore one turn of the cap moves the plug one twenty-fifth of an inch, or ten two hundred-and-fiftieths. The edge of the cap is divided into ten parts, each of which corresponds to the tenth of a complete turn ; and, therefore, to one two-hundred-and-fiftieth of an inch. Hence in reading off the graduations the tens are shown on the body of the whistle, and the units are shown on the edge of the cap.

The scale of the instrument having for its unit the two-hundred-and-fiftieth part of an inch, it follows that the number of vibrations in the note of the whistle is to be found by dividing $\frac{13440 \times 250}{4}$ or 84,000, by the graduations read off on its scale.

A short table is annexed, giving the number of vibrations calculated by this formula, for different depths, bearing in mind that the earlier entries cannot be relied upon unless the whistle has a very minute bore, and consequently a very feeble note.

Scale Readings (one division = $\frac{1}{250}$ of an inch).	Corresponding Number of Vibrations per Second.	Scale Readings (one division = $\frac{1}{250}$ of an inch).	Corresponding Number of Vibrations per Second.
10	84,000	75	11,200
15	56,000	80	10,500
20	42,000	85	9,882
25	33,600	90	9,333
30	28,000	95	8,842
35	24,000	100	8,400
40	21,000	105	8,000
45	18,666	110	7,591
50	16,800	115	7,305
55	15,273	120	7,000
60	14,000	125	6,720
65	12,923	130	6,461
70	12,000		

[1] Mr. Hawksley, surgical instrument maker, 307 Oxford Street, also makes these whistles, and those he makes have much purity of tone.

The largest whistles suitable for experiments on the human ear, have an inner tube of about 0·16 inches in diameter, which is equal to 40 units of the scale. Consequently in these instruments the theory of closed pipes ceases to be trustworthy when the depth of the whistle is less than about 60 units. In short, we cannot be sure of sounding with them a higher note than one of 14,000 vibrations to the second, unless we use tubes of still smaller bore. In some of my experiments I was driven to use very fine tubes indeed, not wider than those little glass tubes that hold the smallest leads for Mordan's pencils. I have tried without much success to produce a note that should be both shrill and powerful, and correspond to a battery of small whistles, by flattening a piece of brass tube, and passing another sheet of brass up it, and thus forming a whistle the whole width of the sheet, but of very small diameter from front to back. It made a powerful note, but not a very pure one. I also constructed an annular whistle by means of three cylinders, one sliding within the other two, and graduated as before.

When the limits of audibility are approached, the sound becomes much fainter, and when that limit is reached, the sound usually gives place to a peculiar sensation, which is not sound but more like dizziness, and which some persons experience to a high degree. Young people hear shriller sounds than older people, and I am told there is a proverb in Dorsetshire, that no agricultural labourer who is more than forty years old, can hear a bat squeak. The power of hearing shrill notes has nothing to do with sharpness of hearing, any more than a wide range of the key-board of a piano has to do with the sound of the individual strings. We all have our limits, and that limit may be quickly found by these whistles in every case. The facility of hearing shrill sounds depends in some degree on the position of the whistle, for it is highest when it is held exactly opposite the opening of the ear. Any roughness in the lining of the auditory canal appears to have a marked effect in checking the transmission of rapid vibrations when they strike the ear obliquely. I myself feel this in a marked degree, and I have long noted the fact in respect to the buzz of a mosquito. I do not hear the mosquito much as it flies about, but when it passes close by my ear I hear a "ping," the suddenness of which is very striking. Mr. Dalby, the aurist, to whom I gave one of these instruments, tells me he uses it for diagnoses. When the power of hearing high notes is wholly lost, the loss is commonly owing to failure in the nerves, but when very deaf people are still able to hear high notes if they are sounded with force, the nerves are usually all right, and the fault lies in the lining of the auditory canal.

E.—QUESTIONS ON VISUALISING AND OTHER ALLIED FACULTIES.

The Questions that I circulated were as follows ; there was an earlier and uncomplete form, which I need not reproduce here.

The object of these Questions is to elicit the degree in which different persons possess the power of seeing images in their mind's eye, and of reviving past sensations.

From inquiries I have already made, it appears that remarkable variations exist both in the strength and in the quality of these faculties, and it is highly probable that a statistical inquiry into them will throw light upon more than one psychological problem.

Before addressing yourself to any of the Questions on the opposite page, think of some definite object—suppose it is your breakfast-table as you sat down to it this morning—and consider carefully the picture that rises before your mind's eye.

1. *Illumination.*—Is the image dim or fairly clear? Is its brightness comparable to that of the actual scene?

2. *Definition.*—Are all the objects pretty well defined at the same time, or is the place of sharpest definition at any one moment more contracted than it is in a real scene?

3. *Colouring.*—Are the colours of the china, of the toast, bread crust, mustard, meat, parsley, or whatever may have been on the table, quite distinct and natural?

4. *Extent of field of view.*—Call up the image of some panoramic view (the walls of your room might suffice), can you force yourself to see mentally a wider range of it than could be taken in by any single glance of the eyes? Can you mentally see more than three faces of a die, or more than one hemisphere of a globe at the same instant of time?

5. *Distance of images.*—Where do mental images appear to be situated? within the head, within the eye-ball, just in front of the eyes, or at a distance corresponding to reality? Can you project an image upon a piece of paper?

6. *Command over images.*—Can you retain a mental picture steadily before the eyes? When you do so, does it grow brighter or dimmer? When the act of retaining it becomes wearisome, in what part of the head or eye-ball is the fatigue felt?

7. *Persons.*—Can you recall with distinctness the features of all near relations and many other persons? Can you at will cause your mental image of any or most of them to sit, stand, or turn slowly round? Can you deliberately seat the image of a well-known person in a chair and see it with enough distinctness to enable you to sketch it leisurely (supposing yourself able to draw)?

8. *Scenery.*—Do you preserve the recollection of scenery with much precision of detail, and do you find pleasure in dwelling on it? Can you easily form mental pictures from the descriptions of scenery that are so frequently met with in novels and books of travel?

9. *Comparison with reality.*—What difference do you perceive between a very vivid mental picture called up in the dark, and a real scene? Have you ever mistaken a mental image for a reality when in health and wide awake?

10. *Numerals and dates.*—Are these invariably associated in your mind with any peculiar mental imagery, whether of written or printed figures, diagrams, or colours? If so, explain fully, and say if you can account for the association?

11.—*Specialities.*—If you happen to have special aptitudes for mechanics, mathematics (either geometry of three dimensions or pure analysis), mental arithmetic, or chess-playing blindfold, please explain fully how far your processes depend on the use of visual images, and how far otherwise?

12. Call up before your imagination the objects specified in the six following paragraphs, numbered A to F, and consider carefully whether your mental representation of them generally, is in each group very faint, faint, fair, good, or vivid and comparable to the actual sensation :—

A. *Light and colour.*—An evenly clouded sky (omitting all landscape), first bright, then gloomy. A thick surrounding haze, first white, then successively blue, yellow, green, and red.

B. *Sound.*—The beat of rain against the window panes, the crack of a whip, a church bell, the hum of bees, the whistle of a railway, the clinking of tea-spoons and saucers, the slam of a door.

C. *Smells.*—Tar, roses, an oil-lamp blown out, hay, violets, a fur coat, gas, tobacco.

D. *Tastes.*—Salt, sugar, lemon juice, raisins, chocolate, currant jelly.

E. *Touch.*—Velvet, silk, soap, gum, sand, dough, a crisp dead leaf, the prick of a pin.

F. *Other sensations.*—Heat, hunger, cold, thirst, fatigue, fever, drowsiness, a bad cold.

13. *Music.*—Have you any aptitude for mentally recalling music, or for imagining it?

14. *At different ages.*—Do you recollect what your powers of visualising, etc., were in childhood? Have they varied much within your recollection?

General remarks.—Supplementary information written here, or on a separate piece of paper, will be acceptable.

INDEX